Molecular Biology

MOLECULAR BASIS OF CLINICAL MEDICINE

General Series Editor

C. Thomas Caskey, MD
Director, Institute for Molecular Genetics
Investigator, Howard Hughes Medical Institute
Baylor College of Medicine
Houston, Texas

Forthcoming titles in the series:
Robert Roberts: Molecular Basis of Cardiology
Emil J. Freireich: Molecular Basis of Oncology

Molecular Basis of Neurology

EDITED BY

P. Michael Conneally, PhD

Distinguished Professor of Medical and Molecular Genetics and
   Neurology
Department of Medical and Molecular Genetics
Indiana University Medical Center
Indianapolis, Indiana

BOSTON

Blackwell Scientific Publications
Oxford   London   Edinburgh   Melbourne
Paris   Berlin   Vienna

Blackwell Scientific Publications

EDITORIAL OFFICES:
238 Main Street, Cambridge, Massachusetts 02142, USA
Osney Mead, Oxford OX2 0EL, England
25 John Street, London WC1N 2BL, England
23 Ainslie Place, Edinburgh EH3 6AJ, Scotland
54 University Street, Carlton, Victoria 3053, Australia
Arnette SA, 2 rue Casimir-Delavigne, 75006 Paris, France
Blackwell-Wissenschaft, Meinekestrasse 4, D-1000 Berlin 15, Germany
Blackwll MZV, Feldgasse 13, A-1238 Vienna, Austria

DISTRIBUTORS:

USA
Blackwell Scientific Publications
238 Main Street
Cambridge, Massachusetts 02142
(Telephone orders: 800-759-6102)

Canada
Times Mirror Professional Publishing
5240 Finch Avenue East
Scarborough, Ontario M1S 5A2
(Telephone orders: 416-298-1588)

Australia
Blackwell Scientific Publications (Australia) Pty Ltd
54 University Street
Carlton, Victoria 3053
(Telephone orders: 03-347-0300)

Outside North America and Australia:
Blackwell Scientific Publications, Ltd.
c/o Marston Book Service, Ltd.
P.O. Box 87
Oxford OX2 0DT
England
(Telephone orders: 44-865-791155)

Typeset by Huron Valley Graphics
Cover & interior design by Joyce C. Weston
Printed and bound by Braun-Brumfield

© 1993 BY BLACKWELL SCIENTIFIC PUBLICATIONS

PRINTED IN THE UNITED STATES OF AMERICA
93 94 95 96 5 4 3 2 1

Library of Congress Cataloging in Publication Data

Molecular basis of neurology / edited by P. Michael Conneally,
    p. cm.—(Molecular basis of clinical medicine)
    Includes bibliographical references.
    ISBN 0-86542-150-1
    1. Nervous system—Diseases—Molecular aspects. 2. Molecular
neurobiology. 3. Nervous system—Diseases—Genetic aspects.
    I. Conneally, P. Michael. II. Series.
    [DNLM: 1. Genetics, Biochemical. 2. Molecular Biology.
3. Nervous System Diseases. WL 100 M7182]
RC347.M62 1992
616.8′0471—dc20
DNLM/DLC
for Library of Congress   91-41446
                                                               CIP

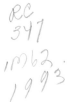

Contents

Contributors

Merrill D. Benson, MD
Professor of Medicine and Medical Genetics, Indiana University, Indianapolis, Indiana

Thomas D. Bird, MD
Chief, Neurology, Veterans Administration Medical Center (Seattle); Professor, Medicine, University of Washington, Seattle, Washington

C. Thomas Caskey, MD
Director, Institute for Molecular Genetics; Investigator, Howard Hughes Medical Institute, Baylor College of Medicine, Houston, Texas

Francis Collins, MD, PhD
Professor of Internal Medicine and Human Genetics, University of Michigan; Investigator, Howard Hughes Medical Institute; Director, Michigan Human Genome Center, Ann Arbor, Michigan

R.C. Eisensmith, PhD
Research Assistant Professor, Department of Cell Biology, Baylor College of Medicine, Houston, Texas

Elizabeth F. Gillard, PhD
Department of Genetics, The Hospital for Sick Children; Department of Molecular and Medical Genetics, University of Toronto, Ontario, Canada

Manuel Rodriguez Gomez, MD
Professor of Pediatric Neurology, Mayo Medical School; Senior Consultant in Neurology, Mayo Clinic, Rochester, Minnesota

James F. Gusella, PhD
Neurogenetics Laboratory, Massachusetts General Hospital, Charlestown, Massachusetts

A.E. Harding, MD, FRCP
Professor of Clinical Neurology, Institute of Neurology, London, England

I.J. Holt, PhD
Department of Biochemical Medicine, Ninewells Hospital and Medical School, Dundee, Scotland

Michele C. LaBuda, PhD
Postdoctoral Fellow, Child Study Center, Yale University School of Medicine, New Haven, Connecticut

David L. Nelson, PhD
Assistant Professor, Institute for Molecular Genetics, Baylor College of Medicine, Houston, Texas

David L. Pauls, PhD
Associate Professor in Genetics and the Child Study Center, Yale University School of Medicine, New Haven, Connecticut

Margaret A. Pericak-Vance, PhD
Associate Medical Research Professor, Duke University Medical Center, Durham, North Carolina

Allen D. Roses, MD
Jefferson-Pilot Corporation Professor of Neurobiology and Neurology; Chief, Division of Neurology, Duke University Medical Center, Durham, North Carolina

Belinda J.F. Rossiter, PhD
Institute for Molecular Genetics, Baylor College of Medicine, Houston, Texas

Moyra Smith, MD, PhD
Professor of Pediatrics, University of California, Irvine, California

Jeffrey Towbin, PhD
Assistant Professor of Pediatrics, Pediatric Cardiology, and The Institute of Molecular Genetics; Phoebe Willingham Muzzy Pediatric Molecular Cardiology Laboratory, Texas Children's Hospital and Baylor College of Medicine, Houston, Texas

Jeffery M. Vance, PhD, MD
Division of Neurology, Department of Medicine, Duke University Medical Center, Durham, North Carolina

Margaret R. Wallace, PhD
Division of Genetics, Department of Pediatrics, University of Florida, Gainesville, Florida

Savio L.C. Woo, PhD
Howard Hughes Medical Institute, Baylor College of Medicine, Houston, Texas

Ronald G. Worton, PhD, FRSC, FCCMG
Department of Genetics, The Hospital for Sick Children, Toronto, Ontario, Canada

Anne B. Young, MD, PhD
Chief, Neurology Service, Massachusetts General Hospital; Julieanne Dorn Professor of Neurology, Harvard Medical School, Boston, Massachusetts

Introduction

The rapid expansion in DNA-based technology has provided new tools for the elucidation and diagnosis of human heritable diseases. The central nervous system (CNS) heritable diseases are difficult to study by conventional biochemical techniques but are very approachable by genetic methods since direct CNS access is not required. The localization of neurological disease genes to regions of the human genome proceeded rapidly following the initial mapping of Huntington's chorea to the short arm of chromosome 4. It is ironic that at the time this text goes to press the Huntington's chorea gene has still not been isolated. The very disease that launched the field remains resistant to solution.

The need for physicians to understand and use the DNA-based technology could not be better illustrated than by the developments related to the discovery of the dystrophin gene and its defects in Duchenne muscular dystrophy and Becker muscular dystrophy. It is now standard practice to diagnose these disorders by rapid and accurate DNA-based methods. This gene was cloned by knowledge of its map position; patients and families now benefit from the research efforts, making it the model for others to follow. The remarkable power of DNA sequence-based diagnosis for Lesch-Nyhan disease points out how rapidly a technique developed in the late 1970s had made its way to the bedside by the late 1980s.

The revolutionary developments of the polymerase chain reaction, and more recently oligonucleotide amplification, now simplify diagnosis of disease genes by selective amplification of the appropriate genetic region. With these advances, the apparently difficult task of DNA-based diagnosis suddenly became simple and transferable to many laboratories.

This text is written for the neurologist who is practicing molecular medicine. It is my opinion that this is an ever-increasing trend in the practice of medicine. Man's estimated 100,000 genes will be known within 15 years as a consequence of the international Human Genome Initiative. Neurology and neuroscience in particular will profit from this knowledge.

Only informed physicians will be able to accomodate the new knowledge for their patients' benefit. This text is written to assist those physicians.

Dr. Michael Conneally has solicited leading researchers in the field of heritable neurological diseases. Each chapter has been formatted to assist the physician in rapid perusal. For those physicians not familiar with the technology and terminology, he has provided an introductory chapter and glossary of terms. Altogether, these elements make this a useful medical text containing the most recent developments.

C. Thomas Caskey, MD

Foreword

We are in the midst of an intellectual revolution in medicine. The advent of molecular biology is transforming medical theory and practice just as, more than a century ago, the advent of the germ theory of disease left an indelible mark. Now, however, the rate of transformation is much faster. There are more scientists, and they start from a more advanced level of knowledge. They have more sophisticated laboratory tools and more effective ways to disseminate information. Each advance in molecular genetics begets more.

One result of these changes is the need, from time to time, to summarize what is happening. Clinicians need the information and so do the investigators themselves. This volume provides a view of the metamorphosis of one speciality, for nowhere in medicine is the impact of molecular genetics more apparent than in neurology. A few examples can be selected from the many provided by the chapters that follow.

Duchenne muscular dystrophy has become the paradigm for a picture of the power of molecular genetics and the problems that remain. Before DNA analysis, the biochemistry of this disease was barely discerned, but the muscle surface membrane was implicated. The affected gene product was not known. As gene mapping was developed, however, the gene was cloned and then the nature of the gene product was deduced, antibodies made, and the protein identified as one hitherto unknown. Armed with that information, a novel approach to gene therapy is being tested, one called "myoblast implantation." Within five years we have gone from virtually total ignorance of the disease to a rational approach to therapy. Even if this approach to treatment is not successful, it will certainly lead to new attempts to replace, modify, or substitute for the missing gene product.

Duchenne muscular dystrophy was already known to be an X-linked disease. In contrast, DNA analysis has led to the recognition of new forms of inheritance. For instance, abnormalities of mitochondrial DNA (mtDNA) have brought recognition of maternal, nonmendelian inheritance, and syndromes that were previously controversial mysteries are being analyzed with new precision. Molecular genetics has not only led to the separation of two major categories of neurofibromatosis, it has done more by helping to explain how tumors arise. Charcot-Marie-Tooth disease was thought to be genetically homogeneous, but now proves to be heterogeneous.

Huntington's disease was the first neurological disease to be mapped (in 1983) and the precise location of the gene is still elusive. Even so, there have been important lessons for human genetics; for instance, homozygous autosomal dominant disease is clinically no more severe than the much more common heterozygous form, and all cases seem to have arisen from the same allele. There has not yet been genetic heterogeneity. Also, attempts to isolate this gene have led to new diagnostic precision and, as a result, presymptomatic diagnosis becomes more and more precise. This, in turn, raises new ethical problems about the identification of individuals doomed to have an untreatable disease. The problem of presymptomatic diagnosis is also a reality for amyloidotic polyneuropathy, a disease in which the gene and gene product are now defined in detail.

Other diseases present different problems and different achievements. Some have given us important lessons for fundamental genetics, such as the nature of mutations in HPRT deficiency, or the analysis of the fragile X syndrome, approaches that were unimaginable a decade ago.

Many challenges remain. Localization of the gene for myotonic muscular dystrophy has proven to be as much a challenge as the Huntington gene. Gene products still have to be identified for many diseases. We have to learn how mitochondrial DNA interacts with nuclear DNA and how changes in mtDNA lead to diverse clinical syndromes. How does any genetic abnormality lead to mental retardation, and why are there so many different forms? Why does lack of dystrophin cause a severe disease in humans, while there are virtually no obvious clinical abnormalities in dystrophin-deficient mice? How does one genetic abnormality lead to pleiotropic clinical syndromes in myotonic muscular dystrophy or tuberous sclerosis?

Young investigators and young physicians have grown up with molecular biology and they are likely to be familiar with the fundamentals. Those who graduated more than a decade ago may need some help with the basics, for which other texts are available. But all of us can benefit from the excellent descriptions of the current state of affairs that are described in the chapters provided by experts in all of these fields.

Lewis P. Rowland, MD
Neurological Institute
Columbia-Presbyterian Medical Center
New York, New York

Preface

In the past decade a veritable explosion of information has occurred in the field of molecular biology. While it affects all areas of medicine, probably its greatest contribution has been in the field of genetic disorders. There are approximately 5,000 such diseases that are inherited in simple mendelian fashion and a host of others in which the mode of inheritance is more complex. While many disorders have been mapped to a specific region of a chromosome, for the majority of these the aberrant gene has not been cloned.

In the case of neurological disorders, which are especially common among these inherited disorders, major research efforts have produced significant advances in the field. The ultimate aim of this research is to find a cure for the disease either by elucidating the basic biochemical defect and then intervening with pharmacologic therapy or by genetic engineering, that is, seeking to correct the defect at the molecular level. This volume portrays these advances, the chapters being written by the leaders in research in their respective diseases. We have attempted to include all of the major neurogenetic disorders. I am deeply grateful to these authors, who took the time to contribute these chapters and also update them with recent findings.

The first chapter, by Dr. Jeffrey Towbin, covers the basic principles of molecular genetics as applied to inherited diseases. The following chapters each begin with the clinical aspects of the disease, followed by the latest development in the molecular approaches to the disorder. While this volume is mainly directed to neurologists and other clinicians and health care workers involved in the management of these disorders, it should also be useful to geneticists and molecular biologists.

I would like to thank Drs. Gail Vance and Belinda Rossiter for their help in editing and ensuring that the authors were consistent in following the guidelines.

P. Michael Conneally, PhD
Indiana University
Medical Center

Glossary of Molecular Genetics

3′ end: See *Ends of DNA fragments.*

5′ end: See *Ends of DNA fragments.*

Allele: One of several alternative forms of a gene that occupies a given locus on the chromosome.

Allele-specific oligodeoxynucleotides (ASO): Synthetic oligodeoxynucleotides prepared to exactly match a sequence flanking and/or including a specific sequence of a gene to be cloned or hybridized. These are usually approximately 20 nucleotides long.

Alternative splicing: A mechanism for generating multiple protein isoforms from a single gene that involves the splicing together of nonconsecutive exons during the processing of some, but not all, transcripts of the gene. A gene is made up of five exons joined by introns. The exons may be spliced by the upper pathway to generate a mature transcript containing all five exons. This type of splicing is termed *constitutive.* The alternative mode of splicing generates a mature transcript that lacks exon 4. If each exon encodes 20 amino acids, the constitutive splicing path would result in a polypeptide made up of 100 amino acids. The alternative path would produce a polypeptide only 80 amino acids long. If the amino acid sequences of the two proteins were determined, the first 60 and the last 20 would be identical. More than 50 genes are known to generate protein diversity through alternative splicing in organisms including *Drosophila,* chickens, rats, mice, and humans.

Alu family: A set of related, highly repeated DNA sequences (each approximately 300 base pairs long) dispersed throughout the human genome; sensitive to digestion by the enzyme *Alu* I.

Anticodon: The triplet of nucleotides in a constant position in the structure of a transfer RNA (tRNA) molecule that associates by complementary base pairing with the codon in the messenger RNA (mRNA) to which the tRNA binds during translation.

Antiparallel: Describes molecules that are parallel but whose internal structures are in the opposite direction, such as the strands of DNA.

Antisense: The DNA strand having the same sequence as messenger RNA (mRNA); there is a nucleotide substitution of thymidine (T) for the uracil (U) found in RNA.

Autoradiography: Detection of images created by the effects of radioactively labeled molecules on x-ray film.

Bacteriophages (phages): Viruses that infect bacteria, including the lambda phage, among others; usually used as cloning vectors.

Base pairs (bp): A partnership of nucleotide bases, such as adenine with thymine or cytosine with guanine, in a DNA double helix held together by hydrogen bonding.

Cap: The structure found at the 5′ end of many eukaryotic mRNAs, consisting of 7′-methyl-guanosine-ppp$X$ (where X is the first nucleotide encoded in the DNA) and added post-transcriptionally near the TATA box.

Centimorgans (cM): Units of recombination representing a recombination fraction of 0.01; defines a constant probability of a crossover per unit length of chromosome and can be used to map genetic distances on chromosomes.

Chromosomal library: Collection of cloned fragments, together representing the DNA of an entire chromosome.

Chromosome walking: Sequential isolation of overlapping DNA clones so as to span large chromosomal intervals.

Cloning: DNA cloning is the insertion of a chosen DNA fragment into a plasmid of a bacterium or a chromosome of a phage, with subsequent replication to form many copies of that DNA.

Coding region of DNA: Gives rise to an RNA molecule of similar sequence (i.e., part of an exon).

Coding region of mRNA: Gives rise to a peptide whose amino acid sequence is determined by the nucleotide sequence of the mRNA.

Codon: A group of three nucleotides that codes for either an amino acid, termination, or an initiation signal.

Cohesive ends: DNA molecules with single-stranded ends that demonstrate complementarity to other ends created through similar means (i.e., cut with the same restriction enzyme) and that can join end-to-end with three fragments.

Complementary DNA (cDNA): Single-stranded DNA copy of a messenger RNA (mRNA) made with the use of the enzyme reverse transcriptase; cDNA contains only the coding sequences of a gene.

Consensus sequence: An average sequence, each nucleotide of which is the most frequent at that position when compared to a set of actual sequences, for example, RNA splice sites, promoter sites, and other sites.

Cosmid Vectors: Plasmids into which lambda phage "cos" sites have been inserted. As a result, the plasmid DNA can be packaged in vitro into the phage head and can be used to infect a bacterium, after which it behaves like a plasmid.

Crossing-over: Exchange of genetic material between chromosomes that pair during meiosis (homologous chromosomes); also called a recombination.

Cytokinesis: Cytoplasmic division, as opposed to karyokinesis.

DNA: Deoxyribonucleic acid, a long polymer of linked nucleotides having deoxyribose as their sugar. They can form double-stranded or helical structures and are the fundamental substance forming genes.

DNA ligase: An enzyme that catalyzes the formation of a phosphodiester bond at the site of a single-strand break in duplex DNA.

DNA polymorphisms: The occurrence of two or more alleles at a given chromosomal locus that may differ by one or several nucleotides.

Dominant inheritance: Provided by an allele that phenotypically manifests in the heterozygous state.

Ends of DNA Fragments (5', 3'): The 5' end refers to the leftward (by convention) or upstream end of the DNA fragment, while the 3' end refers to the rightward or downstream end. Biochemically, 5' and 3' refer to attachment points of phosphate to the ribose ring on the two ends of the coding strand.

Enhancer element: A DNA sequence that facilitates use of some eukaryotic promoters in the *cis* configuration. It can function in any location (upstream or downstream) or orientation, relative to the promoter.

Eukaryote: An organism whose cells contain a nucleus.

Exon: DNA sequences that are transcribed into messenger RNA and represented in the mature RNA product.

Expression vector: Any plasmid or phage in which foreign DNA has been inserted close to a promoter and consequently is transcribed and translated into a protein product.

Gene: A segment of DNA involved in the production of an RNA chain and sometimes a polypeptide. It includes regions preceding and following the coding region, as well as intervening sequences (introns) between coding segments (exons).

Gene locus: The position on a chromosome at which the gene for a particular trait is located; the locus may be occupied by any of the alleles for the gene.

Genetic code: The set of correspondences between nucleotide triplets in DNA and amino acids in proteins.

Genetic linkage: The presence of two or more loci close together on a single chromosome that leads to their inheritance together. Linkage is observed only when the loci are close together since crossing-over usually leads to random assortment of loci that are far apart on the same chromosome.

Genetic polymorphisms: The simultaneous occurrence in the population of genomes showing allelic variation, as seen by the production of different phenotypes or as changes in DNA sequences.

Genome: All the genes of an organism or individual.

Genomic DNA: DNA contained in chromosomes in the nucleus of a cell.

Genomic library: A random collection of DNA fragments obtained from the total genetic material of a cell and carried in a suitable cloning vector.

Genomic probes: Defined nucleic acid segments that can be used to hybridize (and therefore identify) specific DNA clones or fragments bearing the complementary sequence.

Genotype: The genetic constitution of an individual at one or more given loci.

Haplotype: A combination of alleles from closely linked loci, usually with some functional affinity found on a single chromosome.

Heterogeneous nuclear RNA (hnRNA): Comprises the transcripts of nuclear genes made by RNA polymerase II.

Homologous: Refers to structures or processes in different organisms that show a fundamental similarity because of their having descended from a common ancestor. Homologous structures have the same evolutionary origin, although their functions may differ widely (e.g., the flipper of a seal and the wing of a bat).

Homology: Structures with similar morphology. When referring to DNA fragments, it means that they share many nucleotide sequences in common.

Human chromosome band designations: Quinacrine- and Giemsa-stained human metaphase chromosomes show characteristic banding patterns, and standard methods have been adopted to designate the specific patterns displayed by each chromosome. The short (p) arm and the longer (q) arm are each divided into two regions. In the case of the longer autosomes, the q arm may be divided into three or four regions and the p arm into three regions.

Hybridization: The reannealing of single-stranded nucleic acid molecules; the formation of double-stranded regions indicating complementary sequences.

Hydrogen bonding: The formation of weak bonds involving the sharing of an electron with a hydrogen atom. They are important in the specific base pairing in nucleic acids.

Initiation codon: A nucleotide triplet in RNA (AUG) that codes for the first amino acid in protein sequences, formyl-methionine.

Intron (intervening sequences): Any segment of an interrupted gene that is not represented in the mature RNA product. They are part of the primary nuclear transcript but are spliced out to produce mRNA.

Intron–exon junctions: Boundary sequences that separate coding regions from intervening (intron) sequences.

Karyokinesis: Nuclear division, as opposed to cytokinesis.

Kilobases (kb): One thousand nucleotides (or base pairs) in sequence.

Klenow fragment: The large fragment of polymerase I obtained by proteolytic cleavage; it lacks the 5'-to-3' exonuclease activity found in DNA polymerase I.

Library: A set of cloned fragments of DNA, together representing a sample or all of the genome.

Ligases: Enzymes that form C–C, C–S, C–O, and C–N bonds by condensation reactions coupled to ATP cleavage.

Ligation: The formation of a phosphodiester bond to join adjacent nucleotides in the same nucleic acid chain (DNA or RNA).

Linkage: Describes the tendency of genes to be inherited together as a result of their close locations on the same chromosome.

Linker: A small fragment of synthetic DNA that has a restriction site useful for gene splicing (cloning).

lod score: Abbreviation for log odds ratio; is a measure of the confidence in establishing a putative genetic linkage. Numerically this signifies the ratio of the logarithm of the odds for linkage to the logarithm of the odds against linkage at different recombination fractions. A lod score of 3 or more may be interpreted as establishment of linkage.

Messenger RNA (mRNA): An RNA molecule transcribed from the DNA of a gene and from which a protein is translated by the action of ribosomes.

Missense mutant: A mutant in which a codon is mutated to one directing the incorporation of a different amino acid. This substitution may result in an inactive or unstable product.

Nick translation: An in vitro process used to introduce radioactively labeled nucleotides into DNA; utilizes the ability of *Escherichia coli* DNA polymerase I to use a nick in the DNA chain as a starting point from which one strand of duplex DNA can be degraded and replaced by resynthesis with new nucleotides.

Nonsense mutation: A mutation that converts a sense codon to a chain-terminating codon, or vice versa. The results following translation are abnormally short or long polypeptides, generally with altered functional properties.

Northern blotting: A technique for transfer of RNA from an agarose gel to a nitrocellulose or nylon filter on which it can be hybridized to a complementary nucleic acid. It is generally used to examine size and abundance of mRNA.

Nucleosome: The basic subunit of chromatin, consisting of approximately 200 bp of DNA and an octomer of histone protein.

Nucleotide: The portion of nucleic acid composed of a deoxyribose or ribose sugar combined with a phosphate group and nitrogen base (purine or pyrimidine).

Oligonucleotide: A single-stranded linear sequence of nucleotides (typically up to 20 or 30 nucleotides).

Ontogeny: The development of the individual from fertilization to maturity.

Operator: A region of DNA that interacts with a repressor protein to control the expression of an adjacent gene or group of genes.

Palindrome: A DNA sequence that is the same when one strand is read left to right or the other strand is read right to left; frequently a feature of endonuclease recognition sites.

Phenotype: Observable characteristics of an organism resulting from interaction of its genes and the environment in which development occurs.

Phylogeny: The relationships of groups of organisms as reflected by their evolutionary history.

Plasmid: An extrachromosomal, autonomously replicating, circular DNA segment.

Point mutation: 1. In classical genetics, any mutation that is not associated with a cytologically detectable chromosomal aberration or one that has no effect on crossing-over (and therefore is not an inversion) and complements nearby lethals (and therefore is not a deficiency). 2. In molecular genetics, a mutation caused by the substitution of one nucleotide for another.

Polyadenylation: The addition of a sequence of adenine nucleotides to the 3′ end of a eukaryotic RNA after its transcription.

Polymerase: An enzyme that catalyzes the assembly of nucleotides into RNA and of deoxynucleotides into DNA.

Polymerase chain reaction (PCR): An in vitro method for the enzymatic synthesis of specific DNA sequences by repetitive cycles of template denaturation, primer annealing, and extension of annealed primers. This method uses two synthetic oligonucleotide primers flanking the region of interest in the target DNA and DNA polymerase (Taq polymerase) to amplify these sequences.

Polymorphism: The existence of two or more genetically different classes in the same interbreeding population (Rh-positive and Rh-negative humans, for example).

Post-translational modification: Change in chemical structure of a newly formed polypeptide, usually by the addition of glycosyl, sialyl, or amide residues, prior to its use.

Primary transcript: Original unmodified RNA product corresponding to a transcription unit.

primer: A short sequence of DNA (or RNA) that pairs with one strand of DNA and provides a free 3′-hydroxyl group at which DNA polymerase starts the synthesis of a new DNA chain.

Prokaryotes: The superkingdom containing all microorganisms that lack a membrane-bound nucleus containing chromosomes. Cell division involves binary fission. Centrioles, mitotic spindles, and mitochondria are absent.

Aside from pillotinas, prokaryotes also lack microtubules. This super-kingdom contains one kingdom, the Monera.

Promoter: A DNA sequence at which RNA polymerase binds and then initiates transcription; usually includes TATA sequences.

Pseudogene: A gene bearing a close resemblance to a known gene at a different locus, but rendered nonfunctional by additions or deletions in its structure that prevent normal transcription and/or translation. Pseudogenes are usually flanked by direct repeats of 10 to 20 nucleotides; such direct repeats are considered to be a hallmark of DNA insertion.

Purine: A type of nitrogen base; adenine and guanine.

Pyrimidine: A type of nitrogen base; cytosine and thymine.

Reading frame: A nucleotide sequence that starts with an initiation codon, partitions the subsequent nucleotides into amino acid-encoding triplets, and ends with a termination codon. The interval between the start and stop codons is called the *open reading frame* (ORF). If a stop codon occurs soon after the initiation codon, the reading frame is said to be *blocked*.

Recombination: Any process in a diploid cell that generates a new gene or chromosomal combination not found in that cell or its progenitors; at meiosis the process that generates a haploid product whose genotype is different from either of the two haploid genotypes that constituted the meiotic diploid.

Regulatory gene: A gene whose product is involved in the regulation of another gene.

Regulatory sequence: A DNA sequence involved in regulating gene expression (i.e., promoters, enhancers).

Restriction endonuclease: An enzyme that recognizes specific short sequences of (usually) unmethylated DNA and cleaves the duplex. Cleavage is sequence-specific and both DNA strands are cleaved, leaving either blunt or overhanging ends.

Restriction enzyme mapping: Linear array of sites on DNA that are cleaved by various restriction enzymes.

Restriction fragment length polymorphism (RFLP): Inherited variation in the recognition sequences of restriction enzymes that produces different sizes of genomic fragments on Southern blotting.

Retroviruses: RNA viruses that utilize reverse transcriptase during their life cycle. This enzyme allows the viral genome to be transcribed into DNA. The name *retro*virus alludes to this "backward" transcription. The transcribed viral DNA is integrated into the genome of the host cell, where it replicates in unison with the genes of the host chromosome. The cell suffers no damage from this relationship unless the virus carries an oncogene. If it does, the cell will be transformed into a cancer cell. Among the oncogenic retroviruses are those that attack birds (such as the Rous sarcoma virus), rodents (the Maloney and Rauscher leukemia viruses and the mammary tumor agent), carnivores (the feline leukemia and sarcoma viruses), and primates (the simian sarcoma virus). The virus responsible for the current AIDS epidemic is the retrovirus HIV. Retroviruses violate the central dogma during their replication.

Reverse transcriptase: An RNA-dependent DNA polymerase.

RNA splicing: The removal of introns and joining of exons in RNA.

S_1 nuclease mapping: Use of an enzyme (S_1 nuclease) that specifically degrades unpaired (single-stranded) DNA sequences. DNA that is hybridized with RNA is protected and allows identification of the ends of RNA coded by the DNA transcription unit.

Southern blotting: Procedure for transfer of denatured DNA from an agarose gel to a nitrocellulose or nylon filter where it can be annealed with a radiolabeled complementary nucleic acid.

Split gene: A gene that is not continuous but interrupted by intervening sequences.

Tandem repeats: Multiple copies of the same DNA sequence arranged in genes.

TATA (Hogness) box: An A-T–rich sequence (usually 5 bp) found about 25 bp before the start point of each eukaryotic transcription unit using RNA polymerase II. This may also be called a promoter.

Transcription: The formation of RNA from the DNA template.

Transgenic animals: Animals into which cloned genetic material has been experimentally transferred. In the case of laboratory mice, one-celled embryos were injected with plasmid solutions, and some of the transferred sequences were retained throughout embryonic development. Some sequences became integrated into the host genome and were transmitted through the germ line to succeeding generations. A subset of these foreign genes expressed themselves in the offspring.

Translation: The synthesis of protein on an mRNA template.

Vector: An agent (usually a plasmid or phage) consisting of a DNA molecule known to replicate autonomously in a cell to which another DNA segment may be attached experimentally so as to bring about the replication of the attached segment.

Notice

The indications and dosages of all drugs in this book have been recommended in the medical literature and conform to the practices of the general medical community. The medications described do not necessarily have specific approval by the Food and Drug Administration for use in the diseases and dosages for which they are recommended. The package insert for each drug should be consulted for use and dosage as approved by the FDA. Because standards of usage change, it is advisable to keep abreast of revised recommendations, particularly those concerning new drugs.

CHAPTER 1

Principles of Molecular Genetics

Jeffrey A. Towbin

The precursor to present-day molecular genetics has its foundation in the 1968 discovery by S. Linn and W. Arber. Studying extracts of a strain of *Escherichia coli* B cells, they found a specific modification enzyme that methylated unmethylated DNA and a "restriction" nuclease that broke down the unmethylated DNA. Other restriction nucleases and their companion modification methylases were discovered over the next few years, but these were shown to be nonspecific. In 1970, Smith and Wilcox discovered the first true restriction endonuclease in *Haemophilus influenzae* from which it was extracted. This was followed by an explosion of new restriction enzymes found in over 230 bacterial strains, with most recognizing specific groups of four or six bases (Table 1.1). These discoveries, coupled with the subsequent ingenious technical developments provided by E. M. Southern's blotting method, Grunstein and Hogness' colony hybridization isolation technique, and the various sequencing methods, allowed for the rapid growth of molecular genetics as a science.

More recently, Mullis and coworkers at Cetus Corporation developed the polymerase chain reaction (PCR), which revolutionized present-day molecular biology. Molecular genetics has had a major impact on our understanding of the basic science of prokaryotes and eukaryotes (including regulation and bioenergetics), as well as mechanisms of medical disease. The purpose of this chapter is to outline the current state of molecular genetics as it applies to clinical medicine, including present-day methodology and techniques. A glossary of terms is provided in the frontmatter for quick reference; terms in the glossary are in boldface at first use.

Table 1.1:   Some Restriction Enzymes and Their Properties

Microorganism	Abbreviation	Sequence of DNA Cleaved (5′ → 3′) (3′ → 5′)
Escherichia coli RY13	*Eco*RI	↓ GAATTC CTTAAG ↑
Haemophilus influenzae Rd	*Hind*III	↓ AAGCTT TTCGAA ↑
Haemophilus parainfluenzae	*Hpa* I	↓ GTTAAC CAATTG ↑
Haemophilus parainfluenzae	*Hpa* II	↓ CCGG GGCC ↑
Providencia stuartii 164	*Pst* I	↓ CTGCAG GACGTC ↑
Bacillus amyloliquefaciens H	*Bam*HI	↓ GGATCC CCTAGG ↑
Haemophilus aegyptus	*Hae* II	↓ PuGCGCPy PyCGCGPu ↑
Streptomyces albus G	*Sal* I	↓ GTCGAC CAGCTG ↑
Haemophilus influenzae Rd	*Hind*II	↓ GTPyPuAC CAPuPyTG ↑

EUKARYOTIC DNA

In **eukaryotes** (organisms whose cells contain a nucleus), **DNA** generally exists in double strands since the bases tend to form a **hydrogen bond** with each other in a highly specific fashion. Each strand consists of **purine** (adenine, guanine) and **pyrimidine** (cytosine, thymine) bases (or **nucleotides**) linked linearly with phosphodiester bonds. For example, adenine (A) bonds with thymine (T), and guanine (G) bonds with cytosine (C), as a general rule, although mismatches or alternative bonding can occasionally occur. In this way, double-stranded DNA will always contain equimolar proportions of A and T, as well as G and C, but the content of these individual nucleotides will vary widely in DNA from different sources.

The basic unit making up a DNA molecule is called a **base pair** (**bp**), the pair of nucleotides that form hydrogen bonds to each other. This double-stranded DNA molecule conformationally becomes a helix in which the continuous deoxyribosephosphate strands wind around the outside of the helix with the base pairs A+T or G+C in the interior. The base pairs have different bonding and stability characteristics with A+T associating with only two hydrogen bonds (A≡T) and G+C pairing by three hydrogen bonds (G≡C). In regions of the double helix that are rich in A+T residues, the helix can destabilize more easily than in the G+C-rich areas.

The two DNA strands show polarity, with one strand running in the 5′ → 3′ direction, while the complementary strand runs in the opposite direction. This is called an **antiparallel** arrangement. The helix has the ability to adopt multiple conformations. In addition, **genomic DNA** molecules (chromosomes) are very long and can be visualized by electron microscopy. A double helix of 10 bp is 0.34 mm long and 2 nm in diameter. The DNA in the chromosomes exist as linear molecules, and different species usually possess different numbers of chromosomes.

The amount of DNA in cells of different species varies widely, generally with an increase in the DNA content as species complexity increases. In humans, 46 chromosomes are found. With the exception of the sex chromosomes, all other eukaryotic chromosomes are paired, with one partner coming from each parent. Human females have two X chromosomes (XX), while males carry one X chromosome inherited from their mother and a Y chromosome from their father (XY). The two chromosomes of a pair are said to be **homologous** since they will almost always be identical in organization and commonly in the genes carried. However, since there are many mutant genes in a population, a pair of homologous chromosomes may carry genes with slightly different sequences at particular loci and these are called **alleles.**

A mutant gene encoding a defective product can normally be complemented by a good copy of the gene on the homologous chromosome, but if there is a defect on the single copy of the X chromosome that a male carries it cannot be complemented in this way. Thus, there are a large number of sex-linked inherited diseases carried by females but only expressed in males. In reality, however, these disorders can occur in females, but the chances of a female inheriting two defective X chromosomes is quite low.

Since all somatic cells contain a homologous pair of each of the chromosomes, these somatic cells are considered to be diploid (2n). Gametes (ova and sperm), however, contain a single number of each pair of chromosomes and are, therefore, considered haploid (n). The contribution of one parent to the genetic make-up of the offspring is known as the **haplotype.**

In the cell, linear and circular DNA molecules are found in more compact forms. The helix is coiled upon itself multiple times, reducing the overall length significantly, but increasing its diameter. In eukaryotes, this conformation is stabilized by proteins and it allows packing of the DNA into the minimum of space. Upon activation of function, that portion of DNA uncoils and the two helical strands separate transiently.

GENES

Although replication involves the entire DNA molecule, the true functional unit of DNA is a small portion of the molecule, perhaps only hundreds to thousands of base pairs in size. These functional units correspond to the original concepts of genes held by early workers in genetics who postulated these to be associated with a variety of phenotypic characteristics that were inheritable in predictable ways. In any case, the complete collection of **genes** and all other DNA sequences combined in one organism is referred to as the **genome.**

Information used to direct the synthesis of a variety of molecules in highly specific ways is carried by genes. Although the actual number of different nucleotides used to build a DNA molecule is limited to only four, the number of potential arrangements of these nucleotides is vast. For example, a portion of DNA containing 100 nucleotides has 4^{100} (1.7×10^{60}) potential sequences. Since the genetic material of any organism may possess millions of nucleotides, a tremendous diversity of organisms is easily accounted for.

The majority of encoded information contained in the genome is used to direct the sequence of amino acids that are used to form proteins. The expression of this information is mediated by RNA molecules. However, DNA sequences flanking both ends of those that encode information for making RNA also have important regulatory functions for gene activity. It should be noted, however, that a significant portion of the genome does not encode any information.

Repeated Sequences

Perhaps as much as 20–30% of the human genome consists of repetitive sequences of one kind or another. The most abundant of these are the **Alu family** sequences, which all possess a site cleaved by the restriction enzyme *Alu* I and which show strong conservation of sequence. They are approximately 300 bp long and there are $3–5 \times 10^5$ copies in each cell, thus comprising nearly 3% of the human genome, which is 3×10^9 base pairs. These sequences are widely scattered, occurring at intervals of approximately 5,000 bp, and composed of two repeated sequences of 130 bp, one of which has an insert of 31 bp. The human Alu sequences have deoxyadenine runs of nearly 40 nucleotides at the 3′ end, flanked by repeated sequences up to 19 bp long. RNA sequences that hybridize strongly to these Alu sequences are major constituents of **heterogeneous nuclear RNA (hnRNA)**, and are transcribed by RNA polymerase III using promoters located within the transcribed DNA sequences. Thus, the Alu family sequences can be transcribed from any position in the genome.

Human 5S RNA, which is nearly 300 nucleotides long, has a region that is about 80% homologous to the Alu family, and is used to process and remove the leader sequences present on all secreted proteins when they are first synthesized. It is believed that the Alu sequences may be able to move about the genome but, apart from coding for the 5S RNA in humans, have no known functions. They are thought of as "selfish" DNA that does not contribute to the **phenotype** of the organism. This, however, remains controversial.

CONTROL OF THE GENOME

An early and essential event in the expression of genetic information is the binding of RNA polymerase at the start of the **coding regions** or sequences. This specific binding is generally controlled by DNA sequences located 5′ upstream of the coding sequences. These **promoter** sequences, that region to which RNA polymerase binds, are transcribed by RNA polymerase II and include the essential and invariantly located **TATA box** (A-T–rich sequence found approximately 25 bp before the start point of each eukaryotic transcription unit) and a number of additional **promoter/enhancer elements**. These include CCAAT boxes or GC boxes, which are located at variable distances from the TATA site. A higher order of transcriptional regulation is provided by these additional elements. It is clear that DNA sequences important to the **transcription** of eukaryotic genes often lie hundreds or thousands of nucleotides away from the genes, and these sequences are called enhancers. It is believed that these enhancers communicate with promoters by DNA looping that occurs when proteins bind to these sites and to each other. This looping is possibly the mechanism responsible for the action-at-a-distance phenomenon seen in eukaryotic and prokaryotic regulation.

Unlike prokaryotes, there is no direct evidence for the existence of operons in eukaryotic genomes, although some coordinate control of the expression of metabolically related enzymes may occur. Binding of proteins, however, can affect transcription efficiency. This has been demonstrated in cases in which local hormone concentration controls the cellular concentration of a particular **messenger RNA (mRNA)**. In the majority of these instances, the hormone enters the cell, binds to a receptor protein found in the nucleus, and interacts directly with DNA to alter the level of expression of the target gene. Steroid hormones and T_3 (triiodothyronine) enter the cell and act in this way. Polypeptide hormones (e.g., insulin) are thought to exert their actions through release of a second messenger after binding to specific cell membrane receptors. This second messenger then transmits a stimulatory or inhibitory message that controls transcription of certain genes.

Structural Genes and Noncoding Sequences

The total quantity of eukaryotic DNA found per haploid genome is significantly more than that actually required to specify its respective polypeptides. The great majority of this DNA (nearly 99%) is involved in the structural needs of the chromosome while the remainder corresponds to single-copy genes that encode most protein- and gene-specific controlling sequences. Much of the DNA excess is due to repeated nucleotide sequences of various lengths that belong to several classes of DNA, based on copy number per haploid genome. The most highly repetitive sequences are 300 bp long, present in nearly 10^6 copies, and clustered in the heterochromatin. These sequences are not transcribed and are of unknown function. The structural genes contain a class of moderately repetitive sequences that are present in 10^3 copies and vary in size from 1–10 **kilobases (kb)** based on their tendency to reassociate. Reassociation occurs due to **palindromic sequences.** This may result in the formation of the secondary structure of nuclear transcripts. Their function in evolutionary terms appears to be structural rearrangements. A special class of moderately repetitive sequences, the Alu family, appears to be dispersed widely in the eukaryotic genome. Due to their ubiquity, speculations of their function have ranged from organizers of chromatin structure, to initiators of DNA replication, and to recognition sites for tissue-specific transcription.

Another characteristic of the eukaryotic genome that contributes to the DNA excess is the reiteration of genes themselves, that is, some genes are clustered and **tandemly repeated** many times. This may be found in those genes coding for ribosomal RNA (rRNA) and histones, for example, that are organized in tandemly repeated clusters. Untranscribed spacer segments of DNA separate most reiterated genes, but some gene families (e.g., actin) are widely dispersed in the genome. Eukaryotic genes are unique in that nearly all genes are **split,** the protein-coding expressed region (**exons**) being interrupted by noncoding intervening sequences (**introns**) (Figure 1.1). Frequently, introns are longer than exons. This further adds to the size of the genome. Finally, the eukaryotic genome may contain pseudogenes, nonfunctional genes that are homologous to expressed genes but contain altered sequences that prevents them from expressing functional biological products.

The complex nature of eukaryotic gene regulation is also reflected by chromosome structure. Nuclear, genomic DNA is associated with two protein classes—the highly basic histones and various mildly acidic nonhistone proteins—which form the nucleoprotein material called chromatin. The DNA and protein here are organized into repeating units or **nucleosomes,** which contain hundreds of supercoiled base pairs of DNA wound around a histone octomer and are held together by **linker** DNA. The resultant chromatin is a flexible, uninterrupted, unbranched chain of DNA with nucleosome beads formed along its length. The packing of nucleosomes into helical arrays creates the higher-order structure of the chromosome.

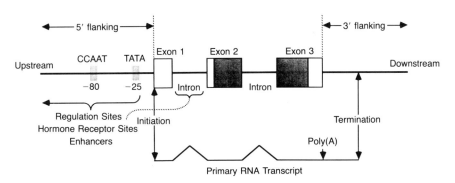

Figure 1.1: General features of a eukaryotic gene. Exons are indicated by boxes, with the shaded regions corresponding to the protein-coding sequences. The unshaded portions at the 5′ and 3′ ends designate the nontranslated mRNA sequences. Regulatory sequences are generally found in the 5′ flanking regions, but enhancers have also been seen.

Transcriptional Control

The vast diversity of cells in humans, all containing identical genomes, indicates that an individual cell expresses only a small part of its full genetic potential. At any given time, only a small fraction of total DNA serves as a template for mRNA synthesis. The following factors appear responsible for gene activation and accessibility for transcription.

Chromatin Conformation   The potential for transcription or expression of a gene is determined primarily by the chromatin conformation of that chromosome. Euchromatin, which corresponds primarily to coding sequences, may be distinguished by its lack of staining by basic dyes, while the staining regions (or heterochromatin regions), contain nontranscribed sequences. Active genes have an altered nucleosome structure and hence a chromosomal domain that is hypersensitive to nucleases. Other sites within this altered domain are hypersensitive to DNase digestion and usually are found flanking the active genes, likely reflecting a relaxed regional DNA configuration lacking in histone H1. In addition, nucleosome alteration occurs via DNA modification, typically methylation. The degree of DNA methylation, primarily at cytosine residues, is inversely related to gene activation. In general, inactive genes are more heavily methylated than active genes. There are, however, cases of specific sites that are methylated more in active genes than in inactive ones. Irrespective of the differences between active and inactive chromatin, the first step in gene activation is likely to be the uncoiling or loosening of tightly packed chromatin.

DNA methylation   5-methylcytosine (5-mC) is the most common DNA modification found in eukaryotic genomes. Approximately 70% of the cytosine residues found in the dinucleotide sequence CpG in humans are methylated in this 5′ position, while other cytosine residues are rarely methylated. Methylation does not affect the ability of cytosine to form a base pair with guanine, so CpG base pairs are formed during replication whether or not the original cytosine residue is methylated. Many of these are maintained in a state through many rounds of cell division.

The CpG dinucleotide is dramatically under-represented in vertebrate genomes, occurring at roughly 25% of the predicted frequency. This "CpG suppression" and the level of DNA methylation appear to be intimately related, probably due to the propensity of cytosine, methylated at the 5′ position, to undergo deamination and form thymidine (i.e., mutation). If true, the long-term effect of the presence of DNA methylation in vertebrate genes would be the gradual loss of CpG dinucleotides and, with them, any functional characteristics that might be bestowed upon that region by these sequences. Youssoufian et al. reported mutations occurring exclusively in CpG dinucleotides in the factor VIII gene causing hemophilia A. This supports the widely accepted theory that the presence of DNA methylation contributes to point mutation. For these mutations to be heritable, they must have occurred in the germ line. Cooper and Youssoufian provided evidence that nearly one-third of intragenic single–base pair mutations causing human inherited disease occur in CpG dinucleotides, the result of C→T or G→A transitions. This hypermutable dinucleotide has been estimated to be at least 40 times more mutable than predicted from random mutation.

In a number of cases where genes are not being transcribed (such as certain tissues or developmental stages) there are methylated cytosine residues in their 5′ flanking regions. In studies where cytosine residues have been methylated upstream of certain normally transcribed genes, these genes are no longer expressed after they are introduced into cells. Methylation of these residues, therefore, appears to provide a means of controlling gene expression. In addition, it is easy to study the degree of methylation of certain sites since the restriction enzyme *Hpa* II only cleaves the sequence CCGG when the cytosines are unmethylated, while *Msp* I cleaves these sequences irrespective of methylation status. However, some cytosines not in this sequence may become methylated and could escape detection.

Regulatory DNA Sequences   The gene consists of a series of exons and introns bound on either side by noncoding regions containing control sequences for initiation and termination of transcription by RNA polymerase. The **intron-exon junctions** have highly conserved **consensus sequences** (idealized sequences in which each position is filled by the base most often found when many actual sequences are compared). Since these are used typically for **RNA splice** sites, this suggests a common splicing mechanism. On either side, adjacent to the genes, are flanking regions containing various **regulatory sequences** and the domains containing DNase-hypersensitive sites. Within the 5′ flanking region (upstream) are at least two elements that are required to promote efficient initiation of transcription by RNA polymerase II—the initiation (**cap**) site and the **TATA box** (located −25 to −31 bp upstream). The TATA box is necessary for recognition of the precise location of transcription initiation, although a few genes lack a TATA box. The promoter region also contains additional sequences located more than 100 bp from the cap site that either enhance gene transcription or regulate transcription in response to effector molecules. Enhancer elements can function in either orientation, whether in front of, within, or downstream from the gene, and they appear to act as bidirectional entry sites for RNA polymerase II (Figure 1.2).

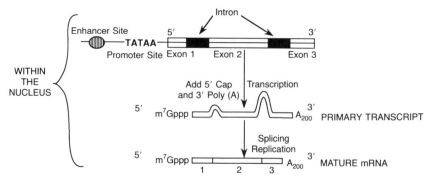

Figure 1.2: Various structures of a eukaryotic gene and its transcription into RNA.

Thus, these are responsible for increased transcriptional efficiency. Short enhancer-like repeated elements with consensus sequences have been noted in the various 5′ flanking regions, occasionally forming **palindromes** with possible regulatory significance. These are unique to a specific set of genes and may convey tissue-specific gene expression.

Although transcription initiation is probably the major control step in gene expression, differential termination of transcription may also be an important factor. The generation of different mRNAs by alternative splicing of the 3′-terminal exons of a gene may depend on the use of different sites for either termination or **polyadenylation.**

Post-Transcriptional Control

This level of control involves RNA processing, a sequence of events used to convert the **primary transcript** of RNA into the mature mRNA, as well as mRNA transport out of the nucleus.

RNA Processing   Primary mRNA transcripts are larger than the mature cytoplasmic counterpart because introns still remain. These transcripts have wide size distribution and for this reason are called heterogeneous nuclear RNA. This processing of hnRNA rapidly initiates after transcription, beginning by enzymatic modifications at both ends of the pre-mRNA chain. Capping of the 5′ end occurs by methylation with the addition of 7-methylguanosine. This 5′ cap appears crucial for mRNA stability and improves ribosome binding. At the 3′ end, after mRNA cleavage, a stretch of approximately 200 adenylate residues is added (polyadenylation), which appears to help mediate subsequent RNA processing and export of mature mRNA from the nucleus (Figure 1.3). In order to convert the primary RNA transcript into an mRNA molecule

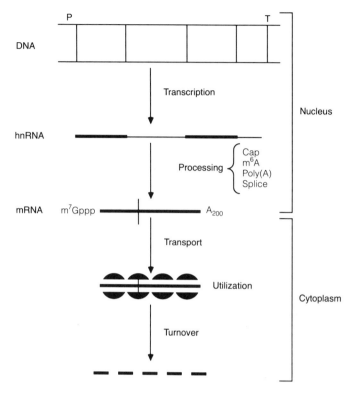

Figure 1.3: Post-transcriptional control could occur at any one of these steps.

coding for a complete protein, introns present in the primary transcript must be removed. The specificity of this splicing mechanism is directed by small nuclear RNAs (snRNAs) containing sequences complementary to consensus splice junctions. In simple processing, the primary transcript is converted into only one mature mRNA by cleavage and ligation at intron-exon junctions (Figure 1.4). In more complex splicing schemes, the primary transcription may give rise to several functionally different mRNAs by the splicing of alternate exon signals. During the splicing process, the majority of the primary RNA transcript mass is removed and degraded, with only 5% of the transcribed RNA being exported to the cytoplasm.

Nucleocytoplasmic Transport RNA is transported through numerous nuclear membrane pores but the exact mechanism is not known. It is presently believed that proteins interacting in the nucleus with RNA, thereby

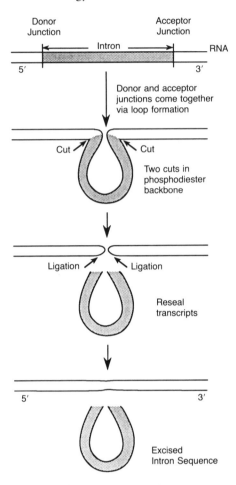

Figure 1.4: Splicing out of introns during the processing of a primary RNA transcript, aided by ribonucleoprotein, leads to the introns forming loops. The chain is then cut at the donor and acceptor junctions and then rejoined.

creating ribonucleoprotein particles (RNP), are important in this transport to the cytoplasm.

Alternate Splicing Mechanisms

The creative molecular mechanisms that are responsible for generation of protein diversity are of two basic types: those that select one gene from a multigene family for expression in a particular cell, developmental stage, or physiological condition (e.g., globin) and those generating a variety of different proteins from a single gene (e.g., troponin T). An example of the latter is alternative RNA splicing, which leads to the differential expression of genomic sequences and the production of multiple protein isoforms from a single gene. While a large number of diverse genes, encoding proteins with myriad functions, utilize alternative splicing, this is particularly prevalent in muscle. Differential splicing has been shown in four of the eight major sarcomeric proteins thus far: myosin heavy chains, alkali myosin light chains, tropomyosin, and troponin T (skeletal and cardiac).

As previously noted, most eukaryotic protein-coding genes contain the sequences found in the resultant mature mRNA as a discontinuous series of DNA segments (exons) interspersed among sequences not later found in the mature mRNA (introns). The primary transcripts of the genes initially contain the intronic sequences before they are excised by the nuclear post-transcriptional regulatory process of splicing. In the majority of instances, each exon present in the gene is incorporated into a mature mRNA via ligation of consecutive pairs of invariant consensus sequences at the 5′ (donor) and 3′ (acceptor) boundaries with removal of all introns. This constitutive splicing process produces a single gene product from each transcriptional unit even when the coding sequence is split into many separated exons. In other instances, however, nonconsecutive exons (or splice sites) are joined in the processing of some gene transcripts, and this alternative pattern of pre-mRNA splicing can exclude individual exons from mature mRNA in some transcripts and include them in others. The use of such differential splicing patterns in transcripts from a single gene creates mRNAs with differing primary structures. When the involved exons contain translated sequences, these alternatively spliced mRNAs will encode related, but distinctly different, protein isoforms. The number of different mRNAs—and therefore protein isoforms—potentially encoded by a given gene increases exponentially as a function of the number of exons participating. This results in a significant increase in phenotypic variability and diversity, all arising from single genes or gene families.

The alternative splicing mechanism appears to work as follows: the primary transcript is capped and polyadenylated to become a suitable substrate and then it complexes with specific ribonucleoproteins to form a "spliceosome." Introns are demarcated by the 5′ donor and 3′ acceptor boundaries, and splicing occurs by donor site cleavage, lariat branch point formation, and cleavage at the acceptor site with concomitant ligation of 5′ and 3′ exons. Small nuclear RNAs have been implicated in this process. Selection of correct pairs of donor and acceptor sites to be joined is imperative, albeit difficult, since splice site sequences (despite being conserved) are repeated elsewhere in the transcript multiple times. Present opinion is that

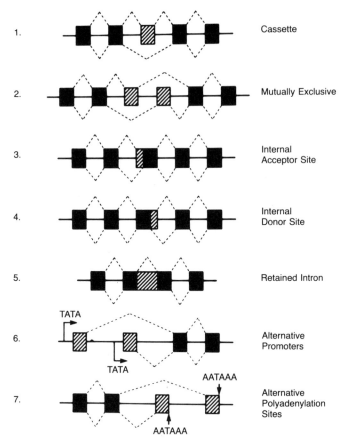

Figure 1.5: Patterns of alternative RNA splicing. Constitutive exons (black), alternative sequences (striped), and introns (solid lines) are spliced according to different pathways (dotted lines), as described in the text. Alternative promoters (TATA) and polyadenylation signals (AATAAA) are indicated. *Reprinted by permission from Breitbant RE, Andreadis A, Nadel-Girard B. Alternative splicing: a ubiquitous mechanism for the generation of multiple protein isoforms from single genes. Ann Rev Biochem 1987;56:467.*

no strict 5′ to 3′ or 3′ to 5′ order exists for intron removal; splicing may instead work through recognition of a characteristic secondary feature by a tracking mechanism rather than intronic primary structure analysis. In addition, alternatively spliced genes may use more than one type of alternative splicing. In theory, a primary gene transcript undergoes alternative RNA splicing if at least one pair of donor and acceptor sites are joined in the formation of another mRNA. Each may remain unspliced or be spliced instead to an alternative partner. At least two mRNAs with different primary sequences result. For example, the rat skeletal muscle troponin T gene consists of several exon types, such as those that are constitutive (which must be included in the mRNA), those that are combinatorial (which can be excluded or included in any combination), and those that are mutually exclusive (where one alternative exon must be included in the mRNA). As a result, up to 64 different mRNAs are produced from this single gene. Those exons not required are spliced out at the same step as the introns.

At this time, seven types of alternative splicing schemes have been described. These include (Figure 1.5):

1. *Combinatorial exons:* A variety of alternatively spliced genes contain entire exons individually included or excluded from the mature mRNA. When these exons are retained, the pattern of splicing is reminiscent of that for constitutive genes in which all potential coding sequences are incorporated into the mature mRNA. When an exon is removed, it is likely carried on a long intron that also contains its flanking noncoding sequences. Such alternatively spliced exons represent discrete *cassettes* of genetic information encoding peptide subsegments that are differentially incorporated into the mature gene product.

2. *Mutually exclusive exons:* In contrast to combinatorial exons in which splicing of each cassette exon appears to be independent of others in the genes, in this pattern one or the other member of a pair is invariably spliced into a given mRNA. Exclusion or inclusion of both members simultaneously does not occur. Each mutually exclusive cassette encodes an alternative version of the same protein domain in two distinct mRNAs. Mutually exclusive exons require a strict directionality of splicing or another safeguard against the joining of one exon of the pair to the common donor to be followed by the joining of the other to the common acceptor.

3/4. *Internal acceptor and donor sites:* Splice sites lying at the boundaries separating mRNA-coding and noncoding sequences delineate cassette exons and, while the exon itself may or may not be incorporated, its flanking introns are invariably excluded in the splicing process. In contrast, genes with internal alternative splice sites exist that lie within a potential coding sequence and in which splicing at these sites results in exclusion of a fraction of the otherwise intact exon.

5. *Retained introns:* This process allows incorporation of intron sequences into mRNA by failing to splice both members of a donor-acceptor pair altogether. The retained intron maintains the intact

translational reading frame, thereby creating a longer exon ("fusion exon").

6/7. *Alternative 3'- and 5'-terminal exons:* In a variety of genes, alternatively spliced mRNAs are associated with different transcripts of the same gene. Heterogeneous sites of transcription initiation and of 3'-end formation result in transcripts with distinct primary structure. Different promoters, as well as different poly(A) sites, may specify alternative 5'- and 3'-terminal exons, respectively. These exons are not the typical cassettes, since each is flanked by a single splice site at its internal boundary alone.

MOLECULAR TECHNIQUES

A variety of methods have been developed to study molecular mechanisms and genes, many of which have subsequently become useful for diagnosing various disease states. The development and application of molecular techniques and recombinant DNA technology to the study of genetic diseases has provided a wealth of information about their molecular basis. Significant progress in understanding the molecular defects responsible for several disorders has occurred and this knowledge subsequently led to the development of DNA-based detection methods. Later, prenatal diagnosis of several disorders by analysis of fetal DNA obtained from amniotic fluid cells or chorionic villi also became possible. The following section will describe the commonly used methods available to the molecular biologist in the search for increasing knowledge about molecular mechanisms and the diseases caused by defects in these mechanisms.

DNA Analysis

The basic principle of most of these methods is based on the molecular **hybridization** of a small amount of a pure, labeled DNA probe to a much larger amount of genomic DNA. The probe and target DNA are initially denatured by boiling, creating single-stranded DNA. This is subsequently allowed to renature into double-stranded, labeled molecules under the desired conditions. The **genomic probe** is allowed to hybridize to only perfectly matching genomic DNA sequences (using appropriate, stringent conditions), while the remaining unhybridized sequences are washed away. The hybrids produced (probe–genomic DNA) are then identified by **autoradiography** (if probes were labeled with ^{32}P) or by staining (if a nonradioactive color producing or fluorescent label is used). This methodology is common for all DNA analysis, therefore these specific techniques require some explanation.

Restriction Endonuclease Cleavage Many bacterial species make these enzymes, which protect the bacteria by degrading any invading DNA molecules. Each enzyme recognizes a specific sequence of four to ten nucleotides in foreign DNA (Table 1.1). There are two major types of **restriction endonucleases** (restriction enzymes). Type I enzymes recognize specific nucleotide sequences but their cleavage sites are nonspecific. Type II enzymes recognize a particular target sequence in a duplex DNA molecule and break polynucleotide chains within that sequence to create discrete DNA fragments of defined length and sequence. These Type II restriction enzymes will cut any length of DNA double helix at their recognition sites into a series of fragments (restriction fragments), each being different in size. The DNA nucleotide sequences recognized by the enzymes are typically palindromic—that is, the nucleotide sequences of the two strands are symmetrical in the recognized region. The two strands of DNA are cut at or near the recognition sequence, often with a staggered cleavage that creates **cohesive ends** which are short and single-stranded at both fragment ends (Figure 1.6). These cohesive or "sticky" ends (DNA molecules with single-stranded nds that show complementarity, making it possible to join end-to-end with introduced fragments) can form complementary base pairs with any other end produced by this same enzyme. Enzymes that create "sticky" ends include *Eco*RI, *Hin*dIII, and a majority of the other restriction enzymes. Another type of restriction endonuclease cuts DNA at their recognition

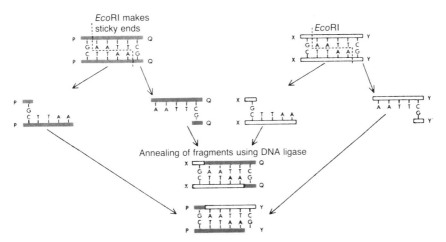

Figure 1.6: The *Eco*RI restriction enzyme makes staggered, symmetrical cuts in DNA away from the center of its recognition site, leaving cohesive or "sticky" ends. A sticky end produced by *Eco*RI digestion can anneal to any other sticky end produced by *Eco*RI cleavage.

sites, creating blunt-ended fragments that are base-paired to their ends (Figure 1.7). These fragments have no tendency to stick together. *Hind*II and *Hpa* I are examples of blunt-end–generating enzymes. Since there are over 100 commercially available enzymes, a large number of different fragments can be generated from DNA stretches.

Agarose Gel Electrophoresis   After the DNA is cleaved by the restriction endonuclease, the digested DNA is loaded into a well of an agarose gel and subjected to electrophoresis (Figure 1.8). The DNA fragments are negatively charged and therefore migrate toward the anode according to their size, with the larger fragments migrating the slowest and remaining nearest to the wells at the top of the gel. After running the gel for sufficient time to separate the fragments, the gel is stained with ethidium bromide, which intercalates between bases and fluoresces under ultraviolet illumination, allowing photographs of the DNA fragments or smear to be obtained. If DNA of known size (such as the lambda **bacteriophage** cut with the enzyme *Hind*III) is placed in one of the wells, the fragment sizes generated in the genomic DNA can be estimated later.

Southern Blotting   In this technique, named after its inventor E. M. Southern, the DNA on the gel is either nicked with acid (HCl, 0.2 Normal) or ultraviolet light to allow for efficient transfer of larger molecular weight fragments, placed in an alkali solution to denature the DNA, then neutralized before being set up for transfer. Until recently, the transfer buffer used was SSC (150 mM NaCl, 15 mM sodium citrate), but many investigators now use sodium hydroxide (NaOH, 0.4 Normal). The transfer of

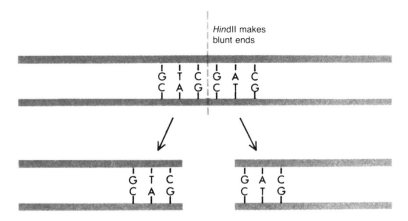

Figure 1.7: The *Hind*II restriction enzyme cuts DNA at the center of its recognition site, leaving blunt ends.

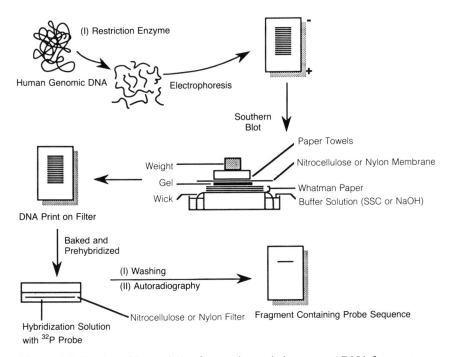

Figure 1.8: Southern blot analysis of genomic restriction enzyme DNA fragments. Initially, the genomic DNA is digested with restriction enzyme, then electrophoresed in an agarose gel. The DNA is then transferred from the gel to a nylon or nitrocellulose filter by capillary flow. This filter is later hybridized to a complementary probe and autoradiographed.

fragments from the gel onto a nylon or nitrocellulose filter occurs by capillary action of buffer flow through the gel (Figure 1.8). The flow of buffer is perpendicular to the direction of electrophoresis and toward the filter. This flow causes the DNA fragments to be carried out of the gel and onto the filter, where they bind and give a replica (or "print") of the DNA fragments previously in the gel. After the transfer, the DNA is fixed to the filter by baking or ultraviolet crosslinking, and then is ready for hybridization to a labeled probe such as a cloned DNA sequence or synthetic oligonucleotide. Although both nitrocellulose and nylon can be reused with different labeled probes after being stripped with detergent, nylon lasts significantly longer.

Hybridization   A variety of methods have been described that are used to label the probe DNA radioactively. The filters to be hybridized are placed in hybridization solution in polyethylene bags that are carefully heat-sealed. The hybridization solutions have classically contained formamide, which decreases the need for high hybridizing temperatures. However, more recently, the method described by Church and Gilbert has been used increasingly. This method employs a simple hybridization solution containing 0.5 M Na_2HPO_4, 1 mM EDTA, and 7% SDS. If the labeled probes are complementary to the DNA of interest, it will anneal (i.e., hybridize) under the proper conditions. The methods of labeling the DNA include the following.

1. *Nick translation:* DNase I introduces nicks at widely separated sites in DNA exposing a free 3'-hydroxyl group that allows DNA polymerase I of *E. coli* to incorporate nucleotides successively. Concomitant hydrolysis of the 5' terminus by the 5' → 3' exonucleolytic activity of polymerase I releases 5' mononucleotides. If the four deoxynucleoside triphosphates (dNTPs) are radiolabeled with ^{32}P, the reaction progressively incorporates the label into a duplex that is unchanged except for translation of a nick along the molecule. Kits are available from several manufacturers.

2. *Oligohexamer or random hexamer labeling:* this labeling scheme produces probes of very high specific activity by denaturing the DNA and then combining random hexadeoxynucleotide primers together with the **Klenow fragment** of DNA polymerase and all four nucleotide triphosphates, one or more of which will be radiolabeled. Klenow fragment, the larger of the two fragments produced when DNA polymerase I is cleaved by subtilisin, retains its 5' → 3' polymerase activity while losing the 5' → 3' exonuclease activity. This enzyme produces a radiolabeled DNA molecule complementary to the nonradioactive denatured DNA. Kits are available from various manufacturers.

3. *Kinase labeling:* this method involves labeling the 5' end of DNA using T4 polynucleotide kinase after the 5' terminus is dephosphorylated. This method is also known as end-labeling and is commonly used to label short oligonucleotides.

Restriction Fragment Length Polymorphism

The discovery of **DNA sequence polymorphisms** by Kan and Dozy has greatly facilitated the genetic analysis of humans. Examination of DNA from any two individuals will reveal DNA sequence variation involving approximately one nucleotide in every 200–500 base pairs. These polymorphic sequences occur much more frequently in DNA than in proteins, and most produce no deleterious clinical effect or destablization of the DNAs.

Some of these DNA sequence changes are detectable by restriction endonuclease digestion of DNA (Figure 1.9). When human DNA from normal individuals is digested with a particular restriction enzyme, fragments of discrete length are obtained. Single–base pair changes may abolish an existing restriction recognition site in the human genome or create a new one, thereby altering the length of these fragments. Alternatively, since the number of tandem repeats interspersed at various intervals in the human genome varies from individual to individual, when these repeat sequences occur between enzyme cleavage sites, the lengths of the DNA fragments generated by the enzyme digestion will vary. **Restriction fragment length polymorphism** (**RFLP**) describes this fragment length variation that is generated by either mechanism (Figure 1.10). The RFLPs have become useful as genetic markers due to their ability to identify the inheritance of

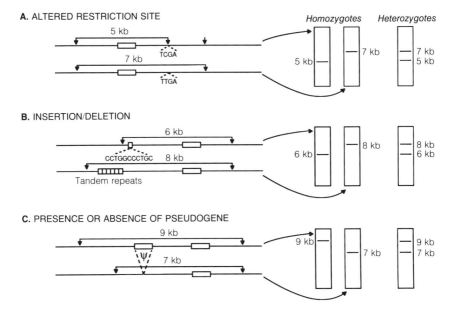

Figure 1.9: Types of DNA polymorphisms. (A) DNA polymorphism resulting from a single-base substitution that eliminates a restriction site and yields a 7-kb rather than a 5-kb fragment (left) that is easily differentiated after Southern blots (right). (B) Insertion/deletion DNA polymorphisms result from a different number of tandem repeats between two restriction sites, which in this case yield 6- or 8-kb fragments. (C) DNA polymorphisms due to presence or absence of a pseudogene change fragment length from 9 kb to 7 kb without affecting the recognition site.

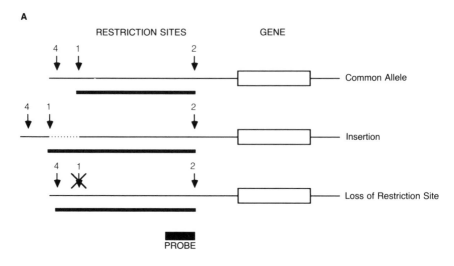

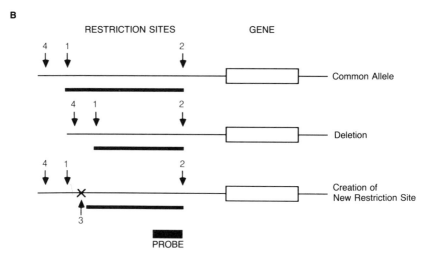

Figure 1.10: (A) Molecular basis of RFLP (smaller fragments). Compared with a common allele, smaller fragments may be generated by deletion of DNA between two restriction sites (here, 1 and 2) or by creation of a new site (here, site 3). *Reprinted by permission from Ostrer H, Hejtmancik JF. Prenatal diagnosis and carrier detection of genetic diseases by analysis of deoxyribonucleic acid. J Pediatrics 1988;112:679–687.* (B) Molecular basis of RFLP (larger fragments). Using a particular probe, a common restriction fragment (or allele) will be found in a population (here, generated by sites 1 and 2). Compared with this allele, larger fragments may be seen. These are shown as bars in the figure. Larger fragments arise from insertion of DNA between two restriction sites (1 and 2) or from loss of site (here, site 1).

the gene of interest. Since polymorphic restriction sites occur frequently on the human genome, a set of such sites can commonly be found in the region of the gene of interest. Performing this analysis to determine whether a set of such sites are present around the gene may allow chromosomes to be classified into different haplotypes as well. These chromosome haplotypes, defined as a combination of alleles from closely linked loci and usually with some functional affinity (alleles segregating together with a particular trait found on a single chromosome) are useful in marking the chromosome at a **gene locus** and also for tracing the origin and migration of genes.

Most of the polymorphic DNA markers for human chromosomes presently available have only two alleles. While two-allele RFLPs can be useful for linkage, they usually are not as informative as systems with many alleles. Fortunately, a small number of known markers (variable number of tandem repeats, VNTRs) detect loci that produce fragments having many different lengths when digested with restriction enzymes. This polymorphism occurs secondary to variations in the number of tandem repeat sequences in that short DNA segment. Since most individuals are heterozygous at these loci, VNTRs can potentially provide linkage information in a large number of families. Nakamura et al. produced a series of single-copy probes from oligomeric sequences derived from tandem repeat regions of a variety of genes and showed them to be highly polymorphic. These sequences have become available for **genetic linkage** analysis and mapping of genetic disease loci.

Synthesis of Polynucleotides

The synthesis of short stretches of nucleotides up to 50 bp long (called **oligonucleotides**) is achieved by organic chemical methods to mRNA sequences. These linear sequences of nucleotides can be used to initiate DNA synthesis on fractions of mRNA, to make a complementary copy (so-called complementary DNA [cDNA]). More recently, oligonucleotides for **polymerase chain reaction** (PCR) or **allele-specific oligonucleotides** (**ASO**) have been prepared from known DNA sequences. If the oligonucleotide (**antisense**) sequence codes for a unique stretch of the peptide of interest, then it may hybridize preferentially with its mRNA to prime the synthesis of cDNA. Another use for oligonucleotides is as probes in the screening of cDNA gene **libraries** in order to identify clones carrying complementary sequences.

Oligonucleotides can be synthesized in cycles, each of which adds one or two bases to the growing chain, with subsequent ligation of these nucleotides. Complex mixtures of oligonucleotides may be created by addition of several nucleotides at certain points in the cycle. Thus, synthesis of a pool of oligonucleotides that represents all possible triplet sequences coding for the amino acids in a particular peptide may be accomplished. This oligonucleotide mixture can then be used as a probe to identify DNA sequences coding for that peptide in a cDNA library and thus allow isolation and identification of the gene.

DNA Sequencing

Two classical methods of DNA sequencing are in common use with the analysis potential of up to 1,000 base pairs daily.

1. *Maxam-Gilbert method:* This chemical method involves isolation of a fragment of DNA labeled at one end with ^{32}P, which is then subjected to a set of four partial, but base-specific, cleavages that produce a series of subfragments. These are separated by size on polyacrylamide urea gels at high voltage, and the labeled fragments are then detected by autoradiography. The base sequence can be read off the sequencing gel autoradiograph from the ladder of each of the base-specific tracks, starting from the bottom of the gel.

2. *Sanger-Coulsen method:* This method involves the **cloning** of the DNA fragment into the single-stranded filamentous virus, M13. Initiation of synthesis of a copy of the inserted DNA whose sequence is desired occurs via a short **DNA primer.** This synthesis is interrupted by four labeled dideoxynucleotides, which terminate growth of the chain at any point where the natural deoxynucleotide should be introduced. The resulting set of products are analyzed in the same gel system as above. Either ^{32}P or ^{35}S can be used.

DNA Cloning and Gene Libraries

DNA fragments from any source can be amplified more than 10^6-fold by inserting them into a **plasmid** (Figure 1.11) or bacteriophage and then growing them in a suitable **vector,** such as bacterial or yeast cells. Plasmids are small, circular molecules of double-stranded DNA naturally occurring in bacteria and yeast, where they independently replicate as the host cell proliferates. Despite generally accounting for only a small percentage of the total DNA in the host cell, they often carry vital genes.

Due to its small size, plasmid DNA can easily be separated from host cell DNA and purified. These purified plasmid-DNA molecules may be used as cloning vectors after being cut by a restriction enzyme, then ligated to the DNA fragment to be cloned. The resultant hybrid plasmid-DNA molecules are then reintroduced into bacteria that have been transiently made permeable to macromolecules, but only a portion of the treated cells take up the plasmid. These cells will survive in the presence of the antibiot-

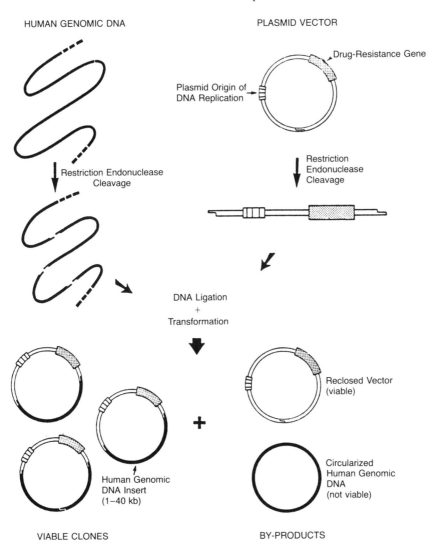

Figure 1.11: Cloning human DNA into a plasmid vector. Human genomic DNA is cleaved with a restriction endonuclease and inserted into a standard plasmid vector that contains a means of replication and a selectable marker (e.g., drug-resistance gene). Various by-products of the procedure (closed vector and circularized human DNA) are also represented.

ic(s) whose resistance genes are encoded by the plasmid. These bacteria divide with concomitant plasmid replication to produce large numbers of copies of the original DNA fragment. The hybrid plasmid-DNA molecules are then purified, and the copies of the original DNA fragment are excised by restriction enzyme digestion.

The cloning process, therefore, may produce millions of different bacterial or yeast colonies, each harboring a plasmid with a different inserted genomic DNA sequence. The rare colony whose plasmid contains the genomic DNA fragment of interest must then be selected and allowed to proliferate and form a large cell population, or clone. Identification of the colony of interest involves use of radioactive nucleic acid probes with sequence complementarity to the desired cloned DNA.

Gene libraries are large collections of individual DNA fragments growing in a suitable host (i.e., *E. coli*). These may be either a **genomic library** (fragments of nuclear DNA), **chromosomal library** (fragments derived from a specific chromosome), or a cDNA library (expressed sequences derived from the total mRNA population of a cell). After growing the library (which may contain more than 10^6 **recombinant** DNA molecules) in the appropriate host on agar plates, the library is transferred to a nitrocellulose or nylon membrane by the technique of replica plating (Figure 1.12). The DNA of the colonies on the filter is denatured by alkali, the filter being baked in an oven with the DNA remaining tightly bound to the membrane and representing an exact copy of the DNA sequences present on the original agarose plates. The membrane is then hybridized with a labeled probe and autoradiographed. The clones that hybridize with the probe are seen as darkened replicas of the clone on the plate. These "positive" clones may be picked and then analyzed by Southern blotting. This method of colony hybridization, first described in 1975 by Grunstein and Hogness, was the first technique that allowed for isolation of cloned DNAs containing a specific gene, and has helped to revolutionize molecular genetics.

The identification of recombinants can be performed using a variety of different probes. These include radiolabeled oligonucleotide probes and cDNA probes (previously described), as well as immunological methods for detection, and expression vectors.

Immunological Methods for Detection   When the protein sequence is unavailable to prepare an oligonucleotide probe, a clone can be identified by use of an antibody to the protein. The DNA bound to the filter is used to hybridize with its corresponding mRNA from a sample of total mRNA. The specific mRNA can be eluted from the filter and translated in an in vitro translation system, the products of which can be identified by

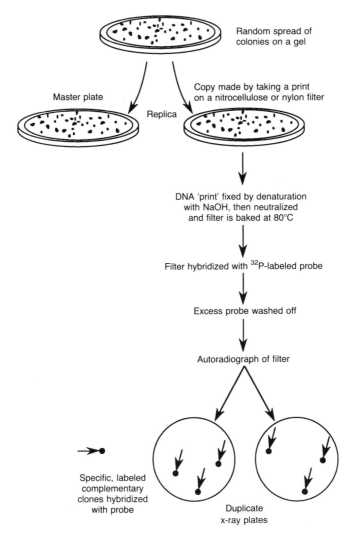

Figure 1.12: Flow diagram for identification of recombinants by hybridization methods. Recombinant clones hybridize with a probe (arrowed clones); the probe can be mRNA, synthetic DNA, or an antibody.

immunoprecipitation. This, then, will allow characterization of the original DNA bound to the filter.

Expression Vectors The vector in this system (e.g., λgt10 or 11) is designed to allow expression of the inserted DNA, so that the host cell will synthesize a part of the normal gene product which then may be detected by immunologic screening methods. A copy of the library is then made and the protein products are firmly fixed by covalent bonding to a membrane. Subsequently, hybridization with the antibody is performed, followed by autoradiographic identification. Positive clones can be selected from the original agar plate and grown to abundance, followed by restriction mapping and DNA sequencing. The DNA sequence can be compared with the amino acid sequence of the protein for confirmation.

Northern Blotting

The Northern blotting method described by Alwine in 1977, is the RNA equivalent of Southern blotting. The RNA is fractionated on an agarose gel and transferred by blotting to nitrocellulose or nylon filters. A radiolabeled DNA probe (usually cloned cDNA or genomic DNA) allows detection of the corresponding RNA sequence bound to the filter. If RNA is extracted from nuclei, the sizes of the precursors can be determined. An estimate of the abundance of the mRNA can be made, and the response to hormonal or metabolic stimuli can be followed. Care must be taken when performing this procedure since RNases are ubiquitous and difficult to eradicate. New glassware or glassware used only for RNA work is generally required.

Polymerase Chain Reaction

The **polymerase chain reaction** (**PCR**) technique, devised by Mullis and coworkers, allows for the specific amplification of discrete DNA fragments. This amplification results in easier detection of nucleic acid fragments initially present in very small (picogram) quantities. Significant reduction in time and labor requirements necessary for the analysis and production of desired DNA fragments occurs since the need for subcloning and plasmid amplification is negated. Isolation and purification of nucleic acid fragments is enhanced because the amplified sequences become the most prominent species in the sample. Existing DNA sequencing methods can be coupled to the PCR technique, directly eliminating the need for cloning and purification of the nucleic acid fragments to be sequenced.

Amplification of DNA sequences by the PCR method mimics the natural DNA replication process of doubling the number of molecules after each cycle (Figure 1.13). The "cycle" here is the repetition of a set of three successive steps performed under different, but controlled, temperatures. The amplified product of interest ("short product") begins to accumulate after as few as three cycles. The short product is the region between the 5' ends of the extension **primers** (synthetic oligonucleotides that anneal to the

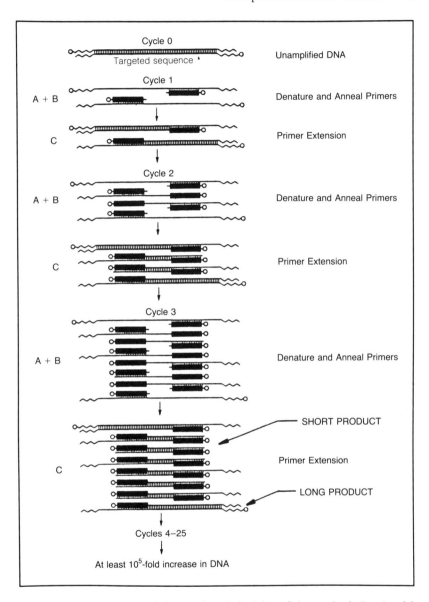

Figure 1.13: Polymerase chain reaction. Principles of the method. *Reprinted by permission from Oste C. Polymerase chain reaction. Biotechniques 1988;6:162–167.*

sites flanking the region to be amplified), which contains discrete ends corresponding to the sequence of these primers. As the cycle number increases, the short product becomes the predominant template to which the extension primers anneal. Theoretically, the amount of amplified product should double after each cycle, leading to exponential accumulation. Due to enzyme kinetics, however, the amount is actually somewhat lower.

As the number of cycles proceeds, other products also form. The "long product," derived directly in each cycle from the template molecules, has variable 3' ends. The quantity of product created increases arithmetically throughout the amplification process since the quantity of original template remains constant. At the end of the amplification process, the short product is typically overwhelmingly more abundant.

The steps that make up the cycles of the PCR method include DNA denaturation, extension primer annealing, and amplification (or extension).

1. *DNA denaturation:* The double-stranded template DNA is denatured under high temperatures and the dissociated single strands remain free in solution.

2. *Extension primer annealing:* Two extension primers, which are selected by the sequence of the DNA at the boundaries of the region to be amplified, are utilized in order to anneal to one of the DNA strands. Each primer anneals to opposite strands. Generally they are different in their sequence and are not complementary to each other. The primers are present in large excess over the DNA template, however, and therefore favor the formation of primer-template complexes at the annealing sites, rather than reassociation of DNA strands when the temperature is lowered.

3. *Amplification (extension):* The 5' → 3' extension of the primer-template complex is mediated by DNA **polymerase** and, as a result, the extension primers become incorporated into the amplification product. Initially, the Klenow fragment of DNA polymerase was used for this reaction but was found to fail occasionally due to the high temperatures required. A thermostable DNA polymerase purified from *Thermus aquaticus,* Taq DNA polymerase, has gained wide usage and has greatly simplified this process since fresh enzyme is not required now after each denaturation step.

Applications of PCR

Cloning   PCR has been utilized successfully for cloning and appears to relieve the usual tedium found in the preparation of DNA fragments seen with classical subcloning methods. Modification of the 5′ ends of the extension primers allows unidirectional cloning into any vector without affecting its ability to anneal specifically to the template. Additional sequences, not complementary to the template, and containing restriction recognition sites, can be attached to the 5′ ends of the extension primers during synthesis and subsequently can be incorporated into the amplified product. This can later be separated from excess primers and dNTPs and digested to generate the ends needed for subcloning.

Chamberlain et al. have recently described a rapid method for scanning megabase regions of the Duchenne muscular dystrophy (*DMD*) gene for deletions utilizing simultaneous genomic DNA amplification of multiple, widely separated sequences. This multiplex genomic DNA amplification procedure was used to amplify specific regions of the *DMD* gene. Failure to amplify a particular region of the gene indicated that the target sequence was not present, i.e., it was deleted. They were able to use this multiplex PCR successfully in several prenatal diagnoses of DMD. This technique appears to have usefulness in deletion detection at any hemizygous locus.

Preparation and analysis of cDNAs may also be enhanced by PCR. Once the first cDNA strand is synthesized, Taq polymerase can be added to promote second strand synthesis. Addition of a pair of specific extension primers allows amplification of specific cDNAs to proceed if the corresponding messenger was present in the initial mRNA. With this method, various tissues can be assayed for expression of the gene.

Frohman and coworkers recently devised a simple and more efficient cDNA cloning strategy for obtaining full-length cDNA clones of low-abundance mRNAs. Using PCR to amplify copies of the region between a single point in the transcript and either end (either 3′ or 5′), cDNAs are generated. The cDNA product may be generated in a single day, Southern blotted and cloned quickly, allowing production of large quantities of full-length cDNA clones of these rare transcripts. To use the rapid amplification of cDNA ends (R.A.C.E.) protocol, a short stretch of sequence from an exon must be known. From this region, primers oriented in the 3′ and 5′ directions are chosen that will produce overlapping cDNAs when fully extended. The primers provide specificity to the amplification step. Extension of the cDNAs from the ends of the messages to the specific primer sequences is accomplished by utilizing primers that anneal to the 3′ end or 5′ poly(A) tail. The overlapping 3′- and 5′-end R.A.C.E. products are combined to produce an intact full-length cDNA. This method provides an efficient alternative to other more time-consuming cDNA cloning methods, as well as potentially being useful in the construction of cDNA libraries.

cDNA cloning using degenerate primers involves the synthesis of oligonucleotide probes based on known amino acid sequence. Prediction of the **codon** usage when designing these probes is difficult since the **genetic code** is degenerate. Lee et al. described the novel synthesis of authentic cDNA probes based on known amino acid sequence, using every possible primer combination coding for the amino acid sequence. The fundamental assumption behind this mixed oligonucleotide primed amplification of cDNA (MOPAC) was that the authentic sequence primer would selectively anneal to its target complementary sequence, out-competing the less complementary primers during the annealing process. They showed that a perfect primer match is not necessary and that there is tolerance of up to 20% base pair mismatch between primer and template during the MOPAC reaction. Mixed primers of more than 1,024 combinations have successfully generated cDNA probes.

Inverse PCR was developed because a major limitation of conventional PCR is that DNA sequences located outside the primer sequences are inaccessible. This is because an oligonucleotide that primes synthesis into a flanking region produces only a linear increase in copy number since no primer in the reverse direction exists. One purpose of inverse PCR is to allow in vitro amplification of DNA flanking a region of known sequence. It utilizes the simple procedures of restriction enzyme digestion of the source DNA and circularization of the cleavage products before amplification using primers synthesized in the opposite orientations to those typically used for PCR. In general, inverse PCR allows for amplification of either upstream and/or downstream regions flanking a specified segment of DNA without resorting to conventional cloning procedures. This method can be used to produce probes rapidly for the identification, as well as the orientation, of adjacent or overlapping clones from a DNA library. This technique eliminates the need to construct and screen DNA libraries to walk thousands of base pairs into flanking regions, and is particularly useful for determining the insertion sites of translocatable genetic elements and other repetitive DNA sequences.

Inverse PCR has been adapted for other applications. One such application is to enzymatically amplify end-specific DNA fragments of a specific orientation from yeast artificial chromosomes (YACs). These probes can then be used for **chromosome walking** in any library containing overlapping DNA fragments and also allows elimination of repetitive sequences from YACs, cosmids, or lambda clones without creating a new construct prior to library screening. Another application was devised by Helmsley et al. where one oligonucleotide primer is synthesized with an alternative base, reflecting the desired modification for site-directed mutagenesis.

Alu PCR was developed to amplify human DNA of unknown sequence from complex mixtures of human and other species DNAs. Previously, application of PCR to isolate and analyze a particular DNA region required knowledge of DNA sequences flanking the region of interest. Initially, it was applied to the isolation of human chromosome fragments in rodent cell backgrounds. This allowed isolation and characterization of sequences from specific regions of human DNA retained in the hybrids, obviating the need for cloned DNA libraries and isolation of human clones through the use of human-specific repeat sequence probes.

Alu PCR has also proved useful for the rapid isolation of human insert DNA from cloned sources, including YACs. This technique utilizes the ubiquitous Alu repeat sequence found in human DNA. There are approximately $3-5 \times 10^5$ copies of this 300-bp sequence distributed throughout the human genome with a known consensus sequence and regions of the repeat that are well conserved. PCR primers designed to recognize these conserved regions allow inter-Alu amplification for isolation of human DNA from complex sources.

Sequencing PCR method: The PCR method may be used as the initial step in sequence analysis, providing the generation of sufficient sample quantities for several subsequent analyses. If the region of interest is first amplified by PCR, the extension primers and dNTPs may be removed and replaced by a third primer (the sequencing primer) that is complementary to one of the strands of the amplified product, followed by the sequencing reaction. This "triple-primer" (Figure 1.14) sequencing method can be performed using the classical Sanger dideoxy sequencing conditions incorporating one radiolabeled deoxynucleotide triphosphate, or alternatively using a third primer radiolabeled at its 5′ end with a radioactive phosphate group.

The triple-primer method requires at least partial knowledge of the sequence being analyzed in order to synthesize the third primer. If no prior sequence information exists, the same procedure may be used provided that the fragment to be sequenced is inserted in a vector of known sequence. Extension primers could be designed to anneal to the vector, resulting in an amplification product whose ends correspond to vector sequences. The third primer could be designed to anneal to the vector, thus initiating the sequencing reaction in the vector and moving into the insert.

Asymmetric PCR: Direct sequencing of the PCR products without an additional cloning step is generally preferred over sequencing cloned products. In addition to its simplicity, it greatly reduces the potential for errors due to imperfect PCR fidelity, as any random misincorporations in an individual template molecule will not be detectable against the greater signals of the "consensus" sequence. Although several reports describing direct sequencing of double-stranded PCR products exist, the protocols

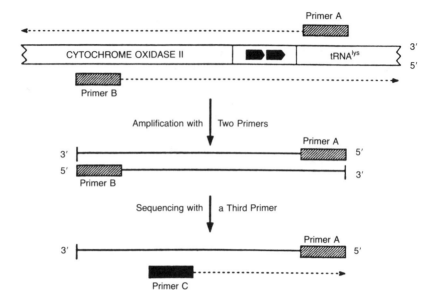

Figure 1.14: Triple-primer sequencing method. The region of interest is first amplified using PCR. Primers A and B are the extension primers. After removing the extension primers and the dNTPs, the third (sequencing) primer C, labeled at its 5′ end with a radioactive phosphate group, is added and the sequencing reaction is conducted in the presence of mixture of deoxy- and dideoxynucleotide triphosphates. *Reprinted by permission from Wrischnik LA, et al. Length mutations in human mitochondrial DNA: direct sequencing of enzymatically amplified DNA. Nucleic Acids Res 1987;15:529–542.*

for preparation of double-stranded template DNA for sequencing were developed for covalently closed circular plasmids.

Difficulties arise when this double-stranded sequencing protocol is applied to PCR-amplified fragments, because of rapid reassociation of the short linear template strands. This can be avoided by modifying the PCR to produce single-stranded DNA of a chosen strand. This modified PCR uses an unequal (i.e., asymmetric) concentration of the two amplification primers.

During the initial 25 cycles, most of the product generated is double-stranded and accumulates exponentially. As the low concentration primer becomes depleted, further cycles generate an excess of one of the two strands, depending on which of the primers was limited. The single-stranded DNA accumulates linearly and is complementary to the limiting primer. The single-stranded template can be sequenced with either the limiting primer or a third, internal primer, which provides an added degree of specificity. This method is less efficient than standard PCR and, therefore, more cycles are generally required to achieve a maximum yield of single-stranded DNA. Usually 30–40 cycles gives the best results.

Mutation detection *Chemical cleavage method:* Cotton and colleagues initially described this technique in 1988 for the study of mutations. Taking advantage of the development of heteroduplexes when thymine (T) is mismatched with cytosine (C), guanine (G), and thymine, or when cytosine is mismatched with thymine, adenine (A), and cytosine, they showed that these heteroduplex DNA sequences, when incubated with either osmium tetroxide (for T and C mismatches) or hydroxylamine (for C mismatches) followed by piperidine incubation, cleaved the DNA at the modified mismatched base. Utilizing end-labeled DNA probes containing T or C single–base pair mismatches, they showed by gel electrophoresis that cleavage was at the base predicted by sequence analysis. This procedure detected all types of mutations, including insertions, deletions, and base changes. Recently, this method has been modified for use in PCR. The speed of PCR and the continued accuracy of this method appears attractive in defining the exact single-base changes occurring in mutated genes.

GC clamps and denaturing gradient gel electrophoresis: Denaturing gradient gel electrophoresis (DGGE) allows separation of DNA molecules differing by as little as a single base change. This separation is based on the melting properties of DNA molecules in solution. DNA molecules melt in discrete segments (melting domains) when the temperature or denaturant concentration is raised. Melting domains vary in size (from 25 base pairs to hundreds of base pairs), each melts cooperatively at a distinct temperature (T_m), and the T_m of a melting domain is highly dependent on its nucleotide sequence due to stacking interactions of the bases. Small sequence changes can lead to large T_m changes, as much as 1.5°C.

In DGGE, DNA fragments are electrophoresed through a polyacrylamide gel containing a linear (top to bottom) gradient of increasing DNA denaturant concentration. As the DNA fragment enters the concentration of denaturant where its lowest T_m exists, the molecule forms a branched structure with retarded mobility in the gel matrix. Hence the fragments separate. DGGE can be used to detect single-base changes in all but the highest T_m of a DNA fragment due to the loss of sequence-dependent migration of fragments upon complete strand dissociation. This can be overcome with cloned DNA fragments by attaching a GC-rich segment—a "GC clamp"—to a DNA fragment that melts in two domains.

In the absence of the GC clamp, only those single-base changes that lie in the first melting domain of this DNA fragment will separate by DGGE, while attaching the GC clamp allows the separation of essentially all single-base changes that lie in the second domain. A 30- to 45-base pair GC clamp can be attached to DNA fragments during the PCR, allowing detection of a single-base change in the attached DNA fragments during DGGE. The huge amplification of DNA fragments by PCR during this procedure increases the sensitivity, so that very small amounts of genomic DNA are required and the signals are detectable by ethidium bromide staining.

Allele-specific oligonucleotides and PCR   This method, described by Saiki and coworkers, is based on hybridization of a probe to amplified material. Here, a synthetic ASO probe (usually 19 to 20 bases long) is used to analyze amplified DNA that was spotted on a solid support, such as nylon or nitrocellulose filter (Figure 1.15). When proper reaction conditions are used, the ASO anneals only to perfectly matched sequences, with single mismatches being sufficient to prevent hybridization. Using high stringency conditions to eliminate any potential for hybridization of partially mismatched probes to the target material, analysis is performed. The signal remaining at the end of the washing steps will reflect the complementarity of sequences of the amplified material. Using PCR, this in vitro amplification method produces greater than a 10^5-fold increase in the amount of target sequence and permits analysis of allelic variation with as little as 1 ng of genomic DNA on dot blots or on Southern blots.

RFLP and haplotype analysis   A haplotype is a combination of alleles at two or more loci on the same chromosome. Due to selective forces or short physical distances between loci, combinations of alleles at different loci on the same chromosome may be inherited as a unit. Previously, analysis of the combinations of alleles on individual chromosomes was feasible only by reconstruction of parental chromosomes from the segregation of alleles in pedigrees. Development of PCR haplotype analysis has made possible the genetic analysis of individual chromosomes without resorting to pedigree analysis. Combinations of alleles along a chromosome can be determined from a small number of sperm or single diploid cells using DNA-based typing of regions amplified in vitro by PCR. This method can also be used to study the recombination frequency between loci that are too close for pedigree analysis to yield statistically reliable recombination rates.

Restriction fragment length polymorphism (RFLP) analysis using PCR has been accomplished by the use of PCR primers from specific regions or repeat sequences to amplify DNA from family members. Most

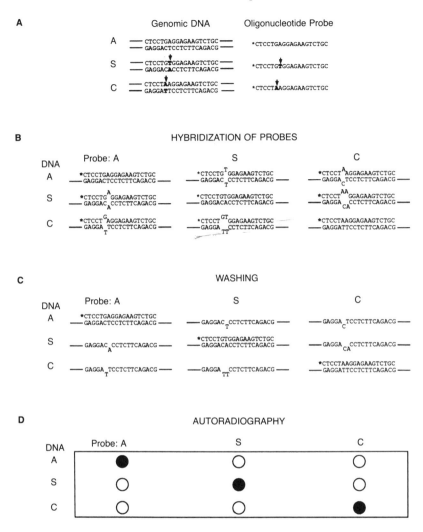

Figure 1.15: Allele-specific oligonucleotide (ASO) screening. (A) DNA sequence around the mutated site in the gene of interest. In this case, the sickle cell anemia mutation site in the β-hemoglobin gene is shown; allele *A* is the wild type, and the mutations in alleles *S* and *C* are indicated with arrows. Each oligonucleotide probe is end labeled as indicated by the asterisk. (B) Using moderately stringent conditions, probes hybridize to all alleles, but where incomplete homology exists, one to two nucleotides mismatch. (C) At higher stringency of washing, only probes matching perfectly remain bound to genomic DNA. (D) Autoradiography detects the presence of bound radiolabeled oligonucleotides and indicates the appropriate allele. *Reprinted by permission from Rossiter BJF, Caskey CT. Molecular studies of human genetic disease. FASEB J 1991;5:21–27.*

of these approaches do not require radioactive labeling, being able to succeed with ethidium-stained gels only. One of the more common uses of PCR RFLP analysis has been that of HLA typing. The HLA region, which is found on the short arm of chromosome 6 (i.e., 6p), encodes a set of highly polymorphic integral membrane proteins that binds antigen peptide fragments. The HLA protein–antigen peptide complex that is formed is recognized by the T-cell receptor, leading to T lymphocyte activation and initiation of a specific immune response. HLA typing, the detection of genetic variation in the HLA region, has traditionally been carried out with serologic reagents or, in the case of class II HLA loci, mixed lymphocyte culture. More recently, HLA typing has been performed at the DNA level by RFLPs generated by Southern blotting and hybridization with cloned HLA cDNA or genomic probes. This time-consuming approach can be replaced by the enzymatic amplification of specific DNA sequences by PCR. The best PCR typing occurs using HLA class II polymorphisms, which include the DP, DQ, and DR series loci. Since these polymorphisms are localized primarily to the N-terminal outer domain encoded by the second exon, PCR primers designed to conserve regions allow for amplification and sequencing of the second exon, resulting in great allelic diversity.

Reverse Genetics

The biochemical bases of the vast majority of human genetic disorders are unfortunately not known, thus making gene mapping and cloning more difficult. After the chromosomal location of the affected gene is established by recombinant DNA and cytogenetic techniques, cloning strategies may be devised to isolate the involved gene. Reverse genetics has been used in the isolation of several disease-causing genes, such as Duchenne muscular dystrophy (*DMD*), retinoblastoma, and cystic fibrosis (*CF*). Linkage analysis, chromosome walking, cDNA analysis, and Southern blotting of the patient's disease-containing genomic DNA are the basis of reverse genetics. Utilizing these techniques, it is possible to learn about both disease and normal processes in cases where only genetic location is known.

Chloramphenicol Acetyltransferase Assays

The chloramphenicol acetyltransferase (CAT) gene is of prokaryotic origin (bacterial transposon Tn9) and has no functional equivalent in eukaryotic cells. Intracellular CAT levels can be readily and sensitively quantitated using commercially available reagents and have been useful in comparisons of transfection efficiency. In addition, since the coding region of this gene can be isolated as a *Sau* IIIa fragment from the plasmid vector pBR325, it can be fused to a eukaryotic promoter/enhancer region and used in the intracellular assays. Measurement of CAT activity provides a rough guide to the level of transcription from the eukaryotic promoter. Deletion mutagenesis of the 5′ and 3′ noncoding regions of this gene fusion can then be performed to establish which domains of the DNA sequence are responsible for modulation of the type of transcriptional regulation under investigation.

A variety of CAT expression vectors have been prepared and used as reporter genes. All vectors contain the bacterial *CAT* gene, followed by an SV40 (simian virus 40) enhancer-promoter region. Putative enhancer elements can be tested in either position or orientation relative to the *CAT* gene, or enhancer-promoter regions under study can be fused to the CAT-coding sequences at the 5′ end of the gene.

Linkage Analysis

Each individual carries two alleles at each locus and transmits one of the two alleles from each locus to an offspring. Generally, the four possible selections of two nonallelic genes, one from each of the two loci, are transmitted in a 1:1:1:1 ratio. However, for certain specific pairs of loci, a deviation from this ratio may be seen because the two nonallelic genes received from one parent tend to be transmitted together to the offspring of this individual (usually due to close physical proximity on the chromosome). This is known as genetic **linkage**. Linkage analysis using RFLPs is widely applied to the detection of genetic disorders (Figure 1.16), and requires loci that are polymorphic (i.e., contain at least two different alleles with appreciable frequency). The more polymorphic the locus, the more useful it is for linkage analysis. One way to measure the degree of polymorphism has been devised by Botstein et al. and has been called the polymorphism information content (PIC). The PIC represents the probability that a given offspring of a random mating between a carrier of a rare dominant gene and noncarrier is informative for linkage between the dominant gene locus and marker locus. When a marker close to or within a defective gene is available, linkage analysis may be used. If the marker and gene are sufficiently close, they will not segregate independently but will be transmitted together. As previously noted, RFLPs result from certain base differences normally seen between homologous chromosomes, and these differences are usually single-base substitutions that occasionally alter the DNA sequence to create or destroy a restriction endonuclease recognition site. Addition or loss of these sites leads to different DNA fragment sizes after endonuclease digestion, with resultant RFLPs that can provide useful linkage analysis markers. Technically, this is performed utilizing the

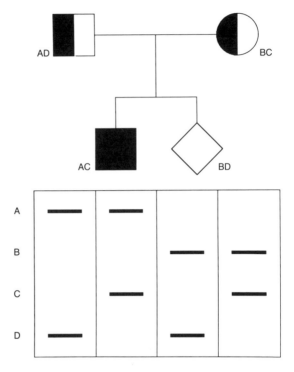

Figure 1.16: Linkage of an autosomal recessive disease gene to a DNA marker by RFLPs. Linked restriction fragments (bands) of different sizes co-segregate either with normal or with disease alleles in family. On the basis of inheritance of the affected child (square), the father's disease allele co-segregates with band A and the mother's with band C. The fetus (indicated by diamond) inherited father's B and mother's D bands, both of which are linked to a normal gene; hence, the fetus is homozygous normal.

procedures previously outlined—digestion with restriction enzymes, followed by agarose gel electrophoresis, Southern blotting, radioactive probe hybridization, and autoradiography. The bands are then analyzed, and the following points should be emphasized:

1. The linkage phase is determined. Linkage phase reflects the relationship between specific alleles of genes and specific alleles of linked markers with respect to chromosome location. For example, couple phase=disease allele and marker allele on the same chromosome; repulsion=disease allele and marker allele on different (homologous) chromosomes. Different linkage phases are seen because many differ-

ent mutations can affect the gene of interest. Therefore, the relationship between the size of the marker and the mutation can differ among families.

2. RFLPs can be used to detect the inheritance of any mutation within a family through this analysis without actually knowing the precise nature of the mutation. For this, samples from close relatives (i.e., affected and normal siblings, spouses, grandparents) are needed to establish the linkage phase between the given mutation and the RFLP being used as the marker.

3. The distance between the RFLP used as a marker and the mutant gene is critical. The greater the distance between the two, the greater the probability of recombination (**crossing over**). Recombination between the gene and the RFLP will cause reversal of the previous linkage phase, resulting in an error in the inferred **genotype**. The map distance between two genes that corresponds to a 1% recombination chance is a **centimorgan** (**cM**). Roughly, 1 million base pairs will result in 1% recombination.

Several methods of linkage analysis are available. The most commonly used method is the likelihood method. This method consists of estimating the recombination fraction (theta) and testing whether an observed estimate is significantly less than 50%. The maximum likelihood (ML) method of estimating recombination fractions is generally used and the likelihood (L)—the probability of occurrence of the phenotypes of all individuals—is calculated. This results in an "odds" ratio (L/0.5), which may be transformed into a decimal logarithm and called the **lod score**. In essence, lod scores are just the logarithm of the odds for linkage versus the odds against linkage. Evidence against linkage occurs when the lod score is highest at theta $= 0.5$. The larger the maximum lod score, the higher the chance for true linkage. Scores of lod $\geq +3$ are considered very significant for linkage, while those ≤ -2 are consistent with nonlinkage. Simply put, a lod score of $+3$ translates to the odds in favor of linkage being 10^3:1 (or 1,000:1). A lod score of -2 is consistent with odds against linkage of 10^{-2}:1 (or 100:1).

Restriction Enzyme Mapping

By restriction endonuclease digestion of DNA utilizing a series of different enzymes, a map of restriction enzyme recognition sites of that DNA may be prepared. This may be used, for example, in defining the introns and exons of a transcript. If both genomic and cDNA clones are available, a comparison of the two may give an approximate indication of the location of exons and introns. This may be confirmed using Southern blot analysis of the genomic DNA with the cDNA.

Chromosome Walking

This technique can be applied to isolate gene sequences whose location is known but whose function is not. First, cloned genomic sequences are localized in the genome by radiolabeling the cloned fragment, then using it as a probe in a hybridization experiment in situ with cytological preparations of chromosomes. In this way, a random set of cloned genomic DNA is localized, and one is chosen whose location on the chromosome in question is closest to the map position of the mutation under investigation. Screening of a genomic library with this clone as a probe ensues, and other overlapping clones that contain the appropriate DNA sequences are selected. The overlap can be on either side (5' or 3') of the probe. Repetition of this step will further select clones with sequences complementary to these overlaps, thus extending the cloned regions. Such walking will occur in both directions along the chromosome unless there is some means of distinguishing the direction (Figure 1.17).

Repetitive DNA sequences are known to be dispersed throughout the genome and these can be troublesome in screening libraries. For this reason, the probe used must be unique in sequence. Once the locus in question is neared, sequence comparison should be possible by direct means or by restriction mapping. Using the walking method, entire genomic sequences together with substantial flanking sequences may be obtained with single steps in each direction.

In Situ Hybridization

A variety of strategies are available for the localization of cloned genes and random DNA sequences within the human genome. In situ hybridization was first described by Gall and Pardue and involves hybridization to a panel of DNA obtained from somatic cell hybrids with different human chromosomes present. For regional localization within a particular chromosome, sequences can be hybridized to the DNA from cell lines carrying deletions of that chromosome, and thereby mapped within or outside the deletion. A panel of somatic cell hybrids with various rearrangements of a particular chromosome can also be used in this way. The smallest region of overlap that gives positive hybridization signals can then be determined. In situ hybridization provides a direct approach to regional mapping by hybridization of known nucleic acid sequences to their complementary DNA within fixed chromosome preparations. Autoradiography is performed

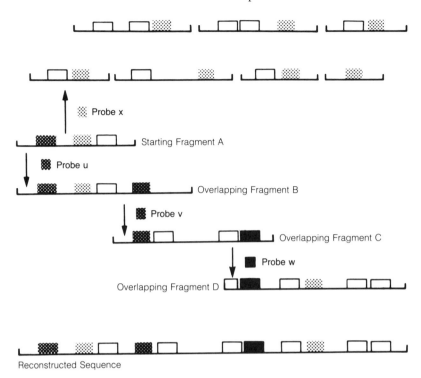

Figure 1.17: Chromosome walking. This figure demonstrates the strategy for reconstruction of the gene order from DNA fragments. Cloned DNA fragments are derived from many cells since typical DNA preparations begin with tissue samples. Therefore, in randomly fragmented DNA preparations, the same sequence is located on overlapping DNA fragments. It is possible to find matching but overlapping fragments, such as fragments A and B. To find the matching sequences, a small radioactive probe (probe u) is prepared from starting DNA fragment A. If the probe is unique, it will hybridize to the unique overlapping fragment B. The unique probe from fragment B (probe v) can then be used to find contiguous fragment C. If the probe is not unique, such as probe x, it will not identify one specific fragment and will not be useful for sequence reconstruction. By repeating these steps with new unique probes from each overlapping sequence, the order of the entire region of the chromosome can be determined. Picture several copies of one manuscript page randomly cut into pieces of paper, each containing five to six words. Finding the same unique word on two different paper fragments would permit reconstruction of the order of two fragments. *Reprinted by permission from Schmickel RD. Contiguous gene syndromes: a component of recognizable syndromes. Pediatr 1986;109:231–241.*

and demonstrates significant excess of silver grains within the hybridized region of a particular chromosome.

This approach has a major advantage over solution hybridization for the quantitation of copy number of message per cell. Hybrid molecules (i.e., DNA-DNA, DNA-RNA, RNA-RNA) formed between the nucleic acids immobilized in cytohistological preparations can also be viewed to obtain information regarding gene expression within a heterogeneous cell population. With this approach, RNA species present in levels as low as 0.01% can be detected. In addition, while examination of stage specificity of mRNA populations in early embryos is difficult by Northern blotting due to insufficient quantities of RNA isolated, in situ techniques can be used to view individual cells to obtain the information needed. Much higher resolution may be obtained by this method for detection of individual RNAs in any cell type at any developmental stage.

In situ hybridization has been used since the early 1970s to localize sequences repeated many times within the human genome. Until the recent DNA recombination technology, the paucity of pure probes stood as an obstacle for localization of other specific gene sequences. The present high-quality probes, together with recent improvements in hybridization efficiency and chromosome banding, have resulted in major improvement in signal resolution. The technique has become sufficiently sensitive to permit localization of single-copy sequence DNA, in addition to the already mentioned repetitive sequences. Thus, this method has become an important complement to the mapping of the human genome.

Site-Directed Mutagenesis

The elucidation of the relationship between the structure and function of genetic material is greatly aided by the availability of suitable mutants. Classical genetic studies selected given prokaryotic phenotypes and examined the genetic alteration in order to obtain mutants. While this is useful for prokaryotes, it is quite difficult to screen for infrequent mutations in higher organisms. The mammalian genome is 1,000-fold more complex than that of bacteria and has long generation times, making classical genetic experiments difficult and time-consuming.

To solve this problem, alterations may be introduced into specific regions of DNA either by point mutation, deletion, or insertion. These mutated molecules can be purified by cloning in bacteria and the alterations in the clones can be characterized by DNA sequencing. The mutated gene may then be tested for activity in a system, in vivo or in vitro, that expresses the normal gene in as physiological a manner as possible.

Site-specific mutagenesis begins with synthesis of an oligonucleotide, usually 15–20 bases long and complementary to the DNA to be altered. A limited internal mismatch, insertion, or deletion is placed in the oligonucleotide, and this mutagenic oligonucleotide is annealed to a single-stranded copy of the DNA (usually cloned into an M13 vector). The complementary strand is synthesized by DNA polymerase, with the oligonucleotide acting as the primer, and results in synthesis of a strand identical to the original parent strand except at nonhomologous positions in the primer. The double-stranded DNA is transformed into E. *coli* and results in the production of two types of bacteriophage—one from the parent strand carrying an unaltered copy of the cloned gene, and the other from the newly synthesized mutated strand.

There are a variety of uses for altered DNA sequences constructed by in vitro mutagenesis, and three general areas can be identified:

1. Investigation of protein structure and function;
2. Studies on the regulatory sites for gene expression and DNA replication; and
3. Modification of therapeutic proteins.

Polymerase chain reaction can now be utilized for in vitro mutagenesis studies. This method allows for increasingly rapid investigation of structure-function relationships.

MEGABASE METHODS

Until the mid-1980s, the size of DNA fragments that could be analyzed or cloned, and the distance that could be covered by chromosome walking, were too small (by one to two orders of magnitude or 20–40 kb) to allow in-depth studies of these complex genomes, a serious limitation of recombinant DNA technology. The total length in base pairs of the human haploid genome is approximately 3×10^9 bp (or 3×10^6 kb) and the smallest distances measurable by recombination analysis is at best of the order of 1 cM (i.e., 1% recombination), corresponding to approximately 1,000 kb. While this is only a mean value, since recombination frequency per length of DNA can vary significantly, it indicates the scale at which genetic analysis operates. For example, cytogenetic analysis, whether based on translocations or deletions, or on in situ hybridization, typically can resolve up to a chromosome band. Since the total number of individualized bands in human chromosomes is approximately 800 bands, the best resolution obtained utilizing this method is of the order of 4,000 kb. As described earlier, "classical" recombinant DNA technology deals with smaller pieces of DNA. Fractionation by agarose gel electrophoresis resolves DNA molecules up to 30–50 kb long, with larger molecules unable to be separated effectively. The effective size range studied by Southern blotting is only a few tens of kilobases around the sequence homologous to the probe utilized. On the other hand, cosmid cloning allows isolation of 40- 45-kb DNA fragments, and chromosome walking strategies based on iterative screening of genomic libraries proceed with "steps" of approximately 20 kb and are both time-consuming and labor intensive.

This large difference between the distance covered by cytogenetics and genetics on the one hand and recombinant DNA techniques on the other led to much frustration. Since many of the problems of human molecular genetics are those requiring movement from point A (the cloned gene or DNA sequence shown by linkage analysis to lie "quite close" to point B (the location of the gene whose dysfunction is responsible for a particular inherited disease), better methods of movement became necessary. This need stemmed from the fact that quite close usually meant 3–5 cM (3,000–5,000 bp) or "in the same chromosome band," well outside the range of conventional blotting, cloning, and even walking techniques. The following methods have recently helped to overcome many of these problems.

Pulsed Field Gel Electrophoresis (PFGE)

This technique, originally developed by Schwartz and Cantor (with later modification by Carle and Olson), allows for improved resolution of large DNA fragments up to 2×10^6 bp (2 megabases). This occurs due to the near linearity of the separation, with particularly good resolution with larger fragment size. In principle, length differences of 10–20 kb are detected with equal ease in fragments ranging from 100–800 kb in size. DNA from entire chromosomes of lower eukaryotes and megabase pair sized DNA from higher eukaryotes have been separated by PFGE.

Given the separation range of PFGE, restriction enzymes must be selected that cleave infrequently. The enzymes with the widest application in normal gel electrophoresis recognize hexanucleotide or smaller sequences and produce fragments averaging 3 kb but ranging from 10 bp to 50 kb. Only a handful of commercial enzymes recognize octameric or longer sequences (i.e., *Not* I, *Sfi* I). Enzymes containing multiple CpG dinucleotides in their recognition site also appear useful since these sequences are under-represented in the genome by one order of magnitude.

For routine DNA analysis, agarose gel electrophoresis of restriction enzyme–digested DNA is the tool of choice. The agarose network functions as a sieve, retaining molecules on the basis of size. Molecules greater than 30 kb cannot be separated by constant electrophoresis. During electrophoresis at high voltages, even small molecules will not size-separate. DNA in solution exists as a free-draining coil and can be thought of as a string of beads. Each bead faces different agarose pores, and migration of DNA in agarose gels requires a single DNA molecule to travel through a series of pores. If different arms of the molecule choose different pores, no net migration occurs. The larger the molecule, the longer it takes for the trailing beads to follow the leading beads through the same network of pores. Since a molecule with fewer beads can choose pores faster than larger molecules, size separation results. Similarly, when a molecule is pulled quickly through agarose (i.e., high voltage), the choice of pores by individual beads is overruled. All molecules will migrate at the same velocity in this instance, independent of size, since the sieving capacity of the agarose has become irrelevant.

PFGE has become useful for the location of genes responsible for genetic defects in man by assisting in the preparation of physical maps of megabase pair DNA regions. Molecular genetic analysis of genomes has been simplified by PFGE.

Yeast Artificial Chromosome (YAC) Cloning

Prior to the development of YAC cloning, the best cloning systems available utilized cosmid vectors. Since the entry of cosmid DNA into the bacterial cells involves λ phage particles, the absolute upper limit to size of the cosmid is the length of DNA that can be packaged in a phage particle (i.e., 50 kb), of which approximately 10–15 kb are used up by the vector sequences. Therefore, the largest DNA fragments that can be cloned by cosmids are only 40–45 kb long. YAC cloning was developed in an attempt to overcome this size limitation, with Burke et al. being the first to successfully implement this system for cloning very large human DNA segments. Development of YACs involves isolating yeast centromere and telomere sequences and combining these with yeast DNA replication origins, selective markers, and plasmid sequences to construct a vector that can be stably propagated by large DNA fragments and introduced into yeast spheroplasts (Figure 1.18).

The 11-kb vector (produced in *E. coli*, where it is propagated as a plasmid) is cleaved by the restriction enzyme *Bam*HI (to remove the "stuffer" fragment) and by *Eco*RI to open up the cloning site. Two "arms" are obtained by partial digestion of high molecular weight DNA with *Eco*RI. The resulting linear DNA molecule contains all the elements needed for maintenance on an artificial chromosome in yeast cells. It is introduced into yeast spheroplasts by transformation using calcium chloride and polyethylene glycol, and the yeast cells containing the artificial chromosome are selected by growth in a medium lacking tryptophan and uracil. Such artificial yeast chromosomes have been shown to allow stable propagation of 100–700 kb of human DNA fragments. This allows for

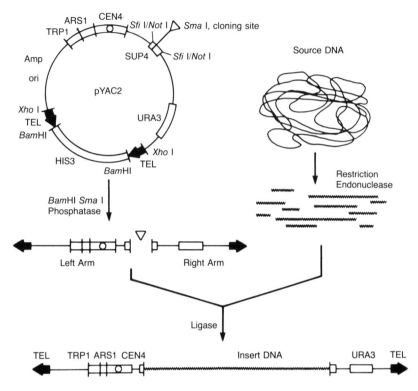

Figure 1.18: Yeast artificial chromosome (YAC) cloning system. The vector, pYAC2, incorporates all necessary functions into a single plasmid that can replicate in *E. coli*, including the cloning site with a gene whose interruption is phenotypically visible (SUP4), an autonomously replicating sequence (ARS1), a centromere (CEN4), selectable markers on both sides of the centromere (TRP1, URA3), two sequences that seed telomere formation in vivo (TEL), and a yeast gene (H1S3) that is discarded during the cloning process. Double digestion of pYAC2 with *BAM*HI and *Sma* I creates three parts, regarded as a left chromosome arm (with centromere), a right chromosome arm, and a throwaway region that separates the two TEL sequences in the circular plasmid. After treating the two arms with alkaline phosphatase to prevent religation, they are ligated onto large insert molecules derived from the source DNA by digesting with an enzyme that leaves blunt ends compatible with *Sma* I–cut ends. The ligation products are then transformed into yeast spheroplasts. *Reprinted by permission from Burke DT, Carle GF, Olson MV. Cloning of large segments of exogenous DNA into yeast by means of artificial chromosome vectors. Science 1987;236:806–812. © American Association for the Advancement of Science.*

the possibility of construction of complete libraries of human DNA in yeast and the cloning of a gene or region of interest in one or a set of such large fragments that can subsequently be mapped in detail or used to prepare mini-libraries. If the mean size of inserts contained in the library is 500 kb, a complete library need only contain 10,000–20,000 clones to cover the genome several times, with each clone able to provide a wealth of information.

Despite the beauty of this system, some technical problems still limit its usefulness. Handling of large DNA fragments in solution, as was reported by Burke et al., is difficult and may limit the maximum insert size. Many investigators are attempting to handle these large DNA fragments in agarose blocks to overcome this problem. In addition, the efficiency of yeast protoplast transformation is still low, giving hundreds of recombinants per microgram of transforming DNA versus the 10^5-10^7 for phage, cosmid or plasmid vectors in *E. coli*. This makes the creation of complete human libraries a task that few laboratories have overcome. The YAC library created by Olson and colleagues, the so-called St. Louis Library, has been distributed to many laboratories for screening to overcome the difficulty of creating these libraries.

The screening of libraries is also less straightforward than for phage, plasmid, or cosmid libraries in *E. coli*. Technically, the mechanics of yeast colony hybridization on filters are similar to the *E. coli* procedure except that prior overnight incubation of filters with zymolase is required to digest the tough cell wall to obtain spheroplasts. Another problem is detection of the yeast clone. In the case of cloning in bacteria, amplification of plasmid or cosmid, or lytic replication of phage, result in colonies or plaques in which the cloned DNA represents a large proportion of the total DNA present. Detection of the signal due to hybridization of a probe is then no problem. Yeast clones, on the other hand, contain just a single copy of the artificial chromosome in the total yeast genome (approximately 15,000 kb), and therefore the signal obtained due to a typical probe 1 kb long may be difficult to detect among background hybridization on all this DNA.

During the past 1–2 years, several laboratories have successfully cloned human genes utilizing YAC libraries. As the number of laboratories utilizing the St. Louis Library, or those developing other libraries, grows the likelihood of this technology leading to the cloning of larger numbers of disease-causing genes increases dramatically.

Chromosome Jumping Libraries

Chromosome walking makes use of overlapping phage or cosmid clones to progressively obtain sequences distal to a given starting point. The problem with this approach lies in the fact that each step, which may extend the existing map of the region of interest by 40 kb (the cosmid insert size) at best, involves screening a complete genomic library, restriction mapping of the resulting clones, and obtaining a new single-copy probe from the end of choice. This makes the approach very labor-intensive. Chromosome walks along several thousand kilobases are, in principle, required to cover the distance between two human loci separated by a few centimorgans, and walks such as these are not feasible. Every sequence between the start point of any walk and the destination, possibly several thousand kilobases away, must be cloned and characterized in this method, but this may be unnecessary if only the point of destination is of interest. Jumping libraries were developed to avoid most of these unnecessary steps and only "touch" the chromosomes at widely spaced intervals (i.e., "jumps"). The basic features of these methods is the circularization of large DNA fragments, which brings together the two ends of the fragment. If performed in the presence of a selective marker, subsequent steps may result in a library in which each clone contains essentially the two ends of the large DNA fragment from which it was derived. In general, jumping libraries should avoid the pitfalls of repetitive regions that plague chromosome walking since these regions can be jumped over.

Two approaches have been developed. Lehrach and colleagues used an approach in which the large DNA fragments are generated by complete digestion of the cellular DNA with a "rare-cutter" enzyme such as *Not* I. The jumping library created contains in each clone the two ends of a given *Not* I fragment. In order to utilize this library, a starting probe is required that is located adjacent to a *Not* I site. To obtain this probe, however, a chromosome walk from the start point to the first *Not* I site may be required. Screening the library with this probe will then pull out a clone containing the other end of the *Not* I fragment, a few hundred kilobases from the starting point. Next, it is necessary to cross the *Not* I site and obtain a probe located on the other side of this site. This can be done by screening a standard **genomic library** or a specially constructed, smaller "junction library" with the probe obtained previously. The procedure can then be repeated, with each jump providing a new probe located several hundred kilobases further from the start point, depending on the length of the particular *Not* I fragment (Figure 1.19).

Two drawbacks are found in the Lehrach complete digest jumping libraries. First, the need for a start probe close to the enzyme site, which adds the necessity for chromosome walking. The second (and more serious) problem is that of missing clones. Very large DNA fragments, (e.g., *Not* I fragments greater than 1,000 kb) will not be easily obtained intact,

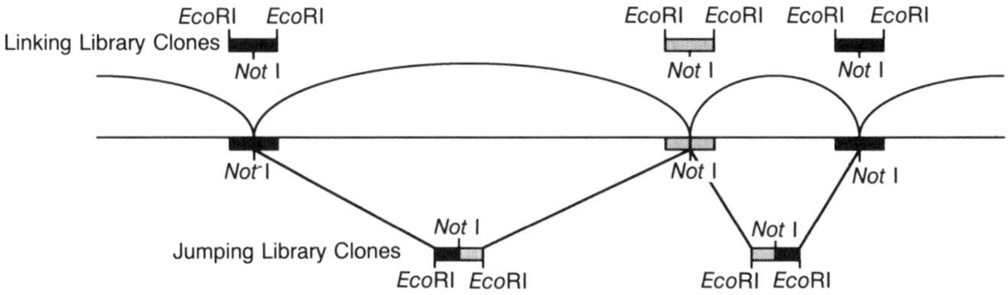

Figure 1.19: Chromosome jumping and linking with *Not* I. Using the "rare-cutter" enzyme *Not* I, large DNA fragments are generated with the ends of each clone having *Not* I ends. These are used to screen the library, pulling out clones containing the other end of the *Not* I fragment, hundreds of kilobases from the starting point. Linking and jumping clones are illustrated here, allowing large distances to be covered rapidly. *Reprinted by permission from Nelson DL. Gene-mapping techniques and applications. Schook, Lewin, McLaren eds. New York: Marcel Dekker, Inc., 1991:65–85.*

and their likelihood of circularizing is much lower than that of smaller fragments. Therefore, this type of jumping library will generally lack clones corresponding to the ends of these very large fragments.

A more general implementation of the jumping library principle was achieved by Collins and coworkers (Figure 1.20). In this case the large DNA fragments are obtained by partial digestion with a frequent-cutter enzyme (i.e., *Mbo* I), and the fragments are subsequently sized by preparative fractionation of pulse field gels before circularization and cloning of the ends as before. The resulting library must be quite large (several million clones) to be representative, but can be used with any starting probes since all sequences should be represented. Since the method is based on circularization of a collection of fragments of similar length, it should not be interrupted by the kind of gap found in Lehrach's method. Jumps of greater than 100 kb are performed with this method. Briefly, the principle of the technique depends on formation of large economic circles from size-selected DNA, bringing the genomic fragments together that initially were far apart. Partial *Mbo* I digestion prior to size-selection is performed, causing no significant bias for a particular sequence to occur at the circle junctions. In each clone, the position of the joining fragments is marked by the *sup* F gene (a suppressor tRNA) and allows for selection of these fragments after restriction enzyme digestion of the circles. These are then ligated into a phage vector. This technique has become a useful and

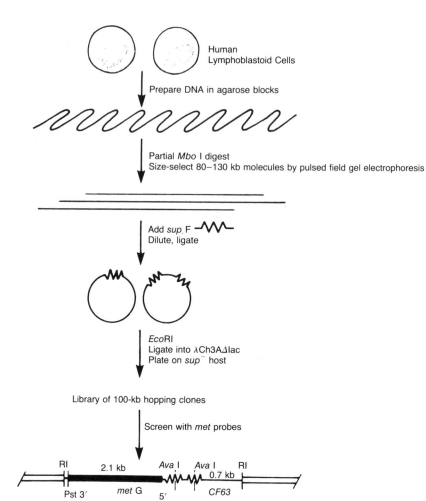

Figure 1.20: This scheme of chromosome jumping was first described by Collins and coworkers. Genomic DNA is circularized and used to generate the jumping library. Each of the phage clones contains two genomic fragments that originally were located at the opposite ends of a 100-kb genomic fragment. Screening of one million phages from the primary library using *met* G resulted in four clones; one is seen at the bottom of the figure. Heavy bars represent the probe utilized for screening (*met* G); the sawtooth mark represents the two *sup* F genes with internal *Ava* I sites that facilitate restriction mapping and subcloning; open bars are phage arms. The 0.7-kb genomic fragment (CF63) should be located approximately 100 kb downstream from the 3′ end of *met* G. *Reprinted by permission from Collins FS, Drumm ML, Cole JL, Lockwood WK, Vande-Woude GF, Iannuzzi MC. Construction of a general human chromosome jumping library, with application to cystic fibrosis. Science 1987;235:1046.* © *The American Association for the Advancement of Science.*

powerful cloning method. Recently, the gene causing cystic fibrosis was cloned with the aid of this technique.

SUMMARY

This chapter has attempted to outline the current state of molecular genetics, including basic concepts and state-of-the-art technology. This area of science has had a dramatic influence on clinical medicine during the past decade and promises to take us ever closer in the quest of understanding disease mechanisms, disease etiologies, and therapy based on disease mechanisms and not its symptoms. In order for medical knowledge and clinical management to keep pace, practitioners will be required to have a solid base in this science. Hopefully, this introduction has given the reader a firm basis upon which the following pages will expand.

ACKNOWLEDGMENTS

Dr. Towbin's work is funded by a grant from the American Heart Association, Texas Affiliate, and a Clinician Investigator Award from the National Institutes of Health.

SELECTED REFERENCES

Alwine JC, Kemp DJ, Stark GR. Method for detection of specific RNAs in agarose gels by transfer to diazobenzylox-methyl paper and hybridization with DNA probes. Proc Natl Acad Sci USA 1977;74:5350.

Botstein D, White R, Skolnick M, David RW. Construction of a genetic linkage map in man using restriction fragment length polymorphisms. Am J Hum Genet 1980;32:314.

Breitbart RE, Andreadis A, Nadal-Ginard B. Alternative splicing: a ubiquitous mechanism for the generation of multiple protein isoforms from single genes. Ann Rev Biochem 1987;56:467.

Burke DT, Carle GF, Olson MV. Cloning of large segments of exogenous DNA into yeast by means of artificial chromosome vectors. Science 1987;236:806.

Collins FS, Drumm ML, Cole JL, Lockwood WK, Vande-Woude GF, Iannuzzi MC. Construction of a general human chromosome jumping library, with application to cystic fibrosis. Science 1987;235:1046.

Grunstein M, Hogness DS. Colony hybridization: a method for the isolation of cloned DNAs that contain a specific gene. Proc Natl Acad Sci USA 1975;72:3961.

Kan YW, Dozy AM. Polymorphism of DNA sequence adjacent to human beta-globin structural gene: relationship to sickle mutation. Proc Natl Acad Sci USA 1978;75:5631.

Maxam AM, Gilbert W. A new method of sequencing DNA. Proc Natl Acad Sci USA 1977;74:560.

Mullis KB, Faloona F. Specific synthesis of DNA in vitro via a polymerase catalyzed chain reaction. Meth Enzymol 1987;155:335.

Nakamura Y, et al. Variable number of tandem repeat (VNTR) markers for human gene mapping. Science 1987;235:1616.

Ott J. Analysis of human genetic linkage. Baltimore: Johns Hopkins University Press, 1985.

Poustka A, Pohl TM, Barlow DP, Frischauf AM, Lehrach H. Construction and use of human chromosome jumping libraries from *Not* I–digested DNA. Nature 1987;325:353.

Saiki RK, et al. Enzymatic amplification of β-globin genomic sequence and restriction site analysis for diagnosis of sickle cell anemia. Science 1985;230:1350.

Sanger F, Coulson AR. A rapid method for determining sequences in DNA by primed synthesis with DNA polymerase. J Mol Biol 1975;94:444.

Schwartz DC, Cantor CR. Separation of yeast chromosome–sized DNAs by pulse field gradient gel electrophoresis. Cell 1984;37:67.

Smith HO, Wilcox KW. A restriction enzyme from *Hemophilus influenzae*. I. Purification and general properties. J Mol Biol 1970;51:379.

Southern EM. Detection of specific sequences among DNA fragments separated by gel electrophoresis. J Mol Biol 1975;98:503.

CHAPTER 2

Hereditary Amyloidosis

Merrill D. Benson

Hereditary amyloidosis is one of the few diseases affecting the nervous system in which genetic mutations responsible for the disease have been characterized. Actually, hereditary amyloidosis is not a single disease since a number of mutations in genes coding for different proteins may cause synthesis of amyloid fibrils. The majority of the hereditary amyloidoses are expressed as peripheral neuropathy and for this reason have long been known as familial amyloidotic polyneuropathy (FAP). Within this group of diseases many have been found to be the result of single-base changes in the gene for plasma transthyretin (prealbumin) with single–amino acid substitutions in the circulating protein. These variant proteins are the building blocks of the amyloid fibrils that define the disease.

Hereditary amyloidosis was first described by Andrade in 1952 when he reported a peculiar form of polyneuropathy in families from northern Portugal. Previous reports of amyloid polyneuropathy, which in retrospect were most likely familial, had not appreciated the hereditary nature of the condition. Shortly after Andrade's original report, Falls et al. described an Indiana/Swiss family in which vitreous amyloid deposits led to blindness. These affected patients also had peripheral neuropathy and restrictive cardiomyopathy. By the end of the 1960s two other kindreds with distinctive patterns of polyneuropathy had been reported, one in Iowa which was associated with a high incidence of peptic ulcer disease and another in Finland in which cranial neuropathy and lattice corneal dystrophy were prominent features. At that point, classification of FAP was made according to ethnic origin and pattern of neuropathy: FAP I—Portuguese, Japanese, and Swedish; FAP II—Indiana/Swiss and Maryland/German kindreds; FAP III—Iowa kindred; and FAP IV—Finnish. While a number of kindreds with FAP have been described subsequently, these four groups have been found to be separate entities, each distinguished by its unique amyloid fibril subunit protein.

Genetically determined forms of amyloidosis without peripheral neuropathy have been described. These include hereditary cerebral hemorrhage with amyloidosis (HCHWA), a condition in which intracranial blood vessels accumulate amyloid deposits which cause progressive and fatal intracranial hemorrhage. Also, approximately 10% of Alzheimer's disease is recognized as being familial. The classic lesion in Alzheimer's disease is the accumulation of amyloid plaques within the central nervous system. Medullary carcinoma of the thyroid, which is part of multiple endocrine adenomatosis (MEA) type II, is associated with amyloid deposits that are composed of a product of the calcitonin gene.

Classification of the Amyloidoses

The term amyloid is a generic one. It refers to proteinaceous material that is deposited extracellularly and that has certain physical chemical characteristics:

1. It is composed of nonbranching fibrils, 75–100 Angstroms (Å) in diameter and having indeterminate length; and
2. It binds Congo red (a planer dye molecule) to give a characteristic green birefringence when viewed by polarization microscopy.

Amyloid deposits contain varying amounts of proteoglycans and also the globular protein serum amyloid P (SAP), but it is the combination of the ordered structure of the fibrils and the ability of the fibrils to bind Congo red that gives the green birefringence. In addition, it is the protein subunit of the fibrils that determines the type of amyloidosis. Classically, there are three types of systemic amyloidosis: (1) immunoglobulin (primary); (2) reactive (secondary); and (3) hereditary (Table 2.1). Recently, a form of amyloidosis with fibrils composed of β_2-microglobulin has been described in patients on long-term hemodialysis. While this β_2-microglobulin amyloidosis mainly affects bones and joints, deposits in other organs have demonstrated the systemic nature of this form of amyloidosis.

In addition to the systemic amyloidoses, a number of localized types of amyloidosis have been described. Few of these have been characterized chemically, presumably due to the limited quantities of material to study.

Pathogenesis of Amyloidosis

Amyloidosis by definition is the extracellular accumulation of protein fibrils that meet the histologic and chemical criteria of amyloid. The patho-

Table 2.1: Systemic Amyloidoses

Type	Chemical Subunit
Immunoglobulin (AL)/Primary	Ig Light Chains
	Kappa
	Lambda
Reactive (AA)/Secondary	Amyloid A
Hereditary (AH)/FAP	Transthyretin
	Apolipoprotein A-I
	Others
β_2-Microglobulin/Dialysis	β_2-Microglobulin

genesis of the amyloid deposits is most likely multifactorial. First, the subunit protein of the fibrils is homogeneous whether it be transthyretin monomers, degradation fragments of serum amyloid A, or immunoglobulin light chain variable segments. Production of the subunit proteins is one variable. In immunoglobulin amyloidosis, a monoclonal gammopathy must be present since all of the subunit protein molecules are of monoclonal nature. The simplest model of amyloidosis is represented by the transthyretin syndromes where single–amino acid substitutions predict the disease (Figure 2.1). Since transthyretin has a preponderance of β structure, it is presumed that this molecule has a strong propensity for β pleated sheet fibril formation. It is then presumed that the single–amino acid substitutions perturb the transthyretin molecule to an extent that causes aggregation or polymerization into fibrils.

A similar situation may be present with the immunoglobulin light chains since considerable β pleated sheet structure is present in these proteins. In this case, partial degradation of the light chain is usually found, and thus it would appear that some degree of biochemical processing is important in fibril genesis. An understanding of the pathogenesis of immunoglobulin amyloidosis of humans has been hindered by the lack of appropriate animal models. In secondary or reactive amyloidosis in the human, only a portion of the precursor protein serum amyloid A (SAA) is found in the fibrils, and, therefore, degradation is of obvious importance. In reactive amyloidosis, the murine model of casein-induced amyloidosis has been used extensively to study generation of SAA protein and its degradation to AA. It has been reported that while in the mouse there are two gene products, SAA1 and SAA2, only SAA2 is incorporated into fibrils. The actual molecular basis for this, however, is not known at the present time.

The pathophysiology of amyloidosis is related to the strategic location

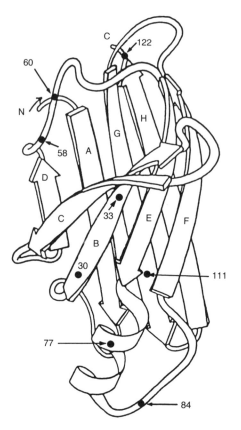

Figure 2.1: Model of the transthyretin (prealbumin) molecule showing eight β pleated sheet strands arranged in two plains. The approximate positions of identified mutations associated with hereditary amyloidosis are noted. While mutations in residues on the outside surface of the molecule might be expected to be associated with aggregation, mutations within the body of the molecule (e.g., positions 30, 33, 111) obviously also predict fibril formation.

and size of the amyloid deposits. Amyloid, in most instances, appears to be an inert substance. It causes pathology by displacement of normal structures (Figure 2.2). This can be appreciated in the restrictive cardiomyopathies, where several hundred grams of the heart mass is amyloid instead of the majority of the mass being functioning muscle fibers. Occasionally amyloid fibril deposits have been observed to diminish in size, but in general, even if the synthesis of the amyloid subunit proteins is stopped, the amyloid deposits will remain.

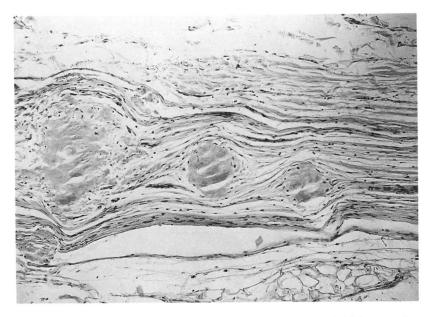

Figure 2.2: Histologic section of peripheral nerve showing amyloid deposits that displace nerve fibers and cause degeneration. Congo red counterstained with hematoxylin.

MENDELIAN HERITABILITY

Practically all forms of hereditary amyloidosis are autosomal dominant. This is as would be expected in a disease where there is a structural gene defect leading to a deposition disease (Figure 2.3). Examination of family pedigrees suggests nearly 100% penetrance in the familial amyloidotic polyneuropathies. This observation may be biased, however, by the fact that families with onset of disease at younger ages have been studied. Now that studies have been done on families with onset of disease in the seventh or eighth decade and direct genomic DNA tests have been developed, it has become obvious that many gene carriers may not develop the disease until very advanced age and perhaps some may not develop it at all. An example of this is the transthyretin methionine-30 variant that causes the classic FAP found in Portugal, Sweden, and Japan. In the families from northern Portugal, the mean age of onset of this syndrome is 32 years for males and 33 years for females. On the other hand, the mean age of disease onset in families from northern Sweden with the same methionine-30

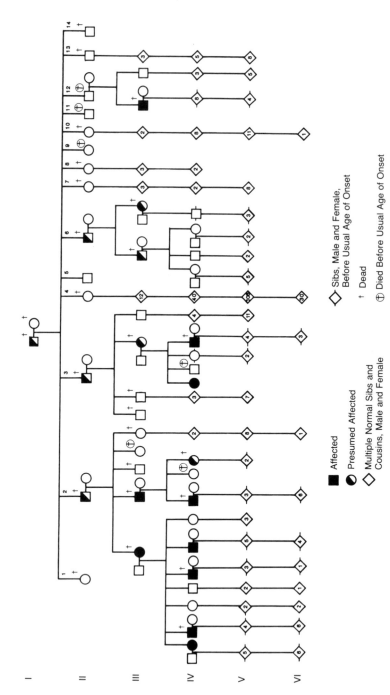

Figure 2.3: Pedigree of a family with hereditary amyloidosis, showing typical autosomal dominant inheritance.

transthyretin mutation is 58 years. Many of the Swedish patients do not manifest peripheral neuropathy and indeed only have vitreous amyloid.

MOLECULAR GENETICS

Hereditary amyloidosis can be the result of mutations in more than one structural gene. Most of the FAP syndromes are related to single–amino acid substitutions in plasma transthyretin (prealbumin) (Figure 2.4). The Iowa kindred with FAP type III, however, has a variant of apolipoprotein A-I. Hereditary cerebral hemorrhage with amyloidosis in Icelandic families is associated with a mutation in the gene for cystatin C. Hereditary cerebral hemorrhage with amyloidosis in a Dutch family is related to the Alzheimer's β amyloid protein.

Transthyretin

Plasma transthyretin, previously known as prealbumin, is the product of a single-copy gene on chromosome 18 (q11.2–q12.1). The gene spans approximately 7 kb and has four exons. The first exon codes for a signal peptide and the first three amino acids of the mature protein (Figure 2.4). No mutations have been described in exon 1. Exon 2 codes for amino acid residues 4 to 47. Two mutations in exon 2 have been verified to be related to hereditary amyloidosis and there is evidence that there are more (Figure 2.5). Exon 3 codes for amino acid residues 48 through 92. Five mutations associated with amyloidosis have been described in exon 3. Exon 4 codes for amino acid residues 93 through 127, the end of the protein, plus approximately 160 bases that are untranslated. The amyloidosis-associated mutations are listed in Table 2.2, which shows the position mutation and amino acid substitution in the mature protein. Transthyretin is expressed mainly in the liver. It has been termed a negative acute phase protein since serum levels decrease in response to inflammation. The exact reasons for this have not yet been determined. Transthyretin is also synthesized in the choroid plexus, which is probably the origin of transthyretin in the cerebral spinal fluid. Transthyretin mRNA has also been detected in the retina and the pancreas.

Transthyretin Amyloidosis Syndromes

Most of the hereditary amyloidoses associated with transthyretin mutations are manifested as peripheral neuropathy. Varying degrees of cardiac, renal, gastrointestinal, ocular, and other organ system involvement may occur with each of the genetic mutations, and indeed the disease in different kindreds with the same mutation may vary considerably. A brief summary of the clinical features of each of the transthyretin amyloidoses follows.

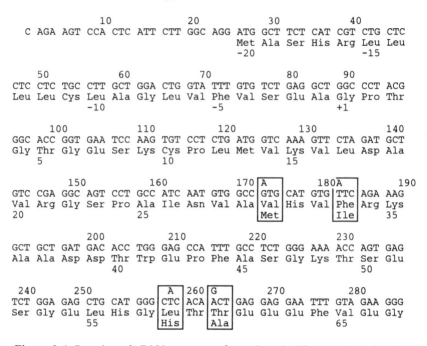

```
            10                20                30                40
  C AGA AGT CCA CTC ATT CTT GGC AGG ATG GCT TCT CAT CGT CTG CTC
                                    Met Ala Ser His Arg Leu Leu
                                    -20                 -15

       50                60                70                80        90
CTC CTC TGC CTT GCT GGA CTG GTA TTT GTG TCT GAG GCT GGC CCT ACG
Leu Leu Cys Leu Ala Gly Leu Val Phe Val Ser Glu Ala Gly Pro Thr
            -10                 -5                      +1

          100               110               120               130               140
GGC ACC GGT GAA TCC AAG TGT CCT CTG ATG GTC AAA GTT CTA GAT GCT
Gly Thr Gly Glu Ser Lys Cys Pro Leu Met Val Lys Val Leu Asp Ala
    5                       10                15

             150               160               170 A           180 A       190
GTC CGA GGC AGT CCT GCC ATC AAT GTG GCC GTG CAT GTG TTC AGA AAG
Val Arg Gly Ser Pro Ala Ile Asn Val Ala Val His Val Phe Arg Lys
20                    25                      Met         Ile     35

                200               210               220               230
GCT GCT GAT GAC ACC TGG GAG CCA TTT GCC TCT GGG AAA ACC AGT GAG
Ala Ala Asp Asp Thr Trp Glu Pro Phe Ala Ser Gly Lys Thr Ser Glu
                40                      45                      50

240       250                   A   260 G     270               280
TCT GGA GAG CTG CAT GGG CTC ACA ACT GAG GAG GAA TTT GTA GAA GGG
Ser Gly Glu Leu His Gly Leu Thr Thr Glu Glu Glu Phe Val Glu Gly
                55      His     Ala                     65
```

Figure 2.4: Protein and cDNA sequence of transthyretin. The mutations that have been shown to be associated with hereditary amyloidosis are indicated by boxes.

Methionine-30   Methionine-30 transthyretin amyloidosis is the most common type of FAP. While first described in families in northern Portugal, it is also present in Japan, Sweden, Greece, Cypress, Majorca, Brazil, Turkey, and the United States. It is almost always associated with a peripheral neuropathy starting in the lower extremities and progressing cephalad. Upper extremities are involved later in the disease, and cranial nerves may also be involved. The first sign of the disease may be bowel dysfunction and/or impotence. Amyloid deposits in the vitreous of the eye occur more frequently in the Swedish kindreds than in other ethnic groups. Scalloped pupil deformity may also occur as a result of a neuropathy of the ciliary nerves. Varying degrees of renal or cardiac amyloidosis may be the cause of death, but many patients die from malnutrition related to gastrointestinal amyloidosis. The mean age of onset of the clinical syndrome in Portuguese patients is 32 for males and 33 for females. In the Swedish kindreds, the mean age of onset is 58, and in patients of English origin the disease may not be manifest until the mid 60s.

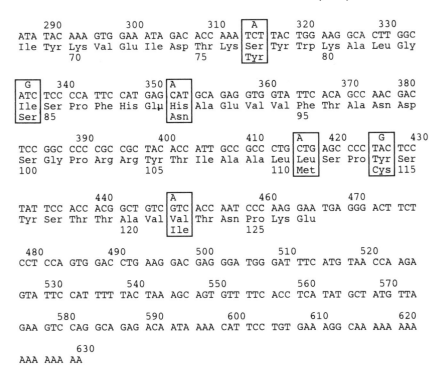

Figure 2.4, continued

Isoleucine-33   Amyloidosis associated with a mutation at position 33 of the transthyretin gene (Phe to Ile) was described in a man who was born in Poland and immigrated to Israel. The neuropathy was described as fairly typical FAP I, in other words similar to the Portuguese syndrome. Diarrhea, sexual impotence, and vitreous amyloid deposits are also features of this syndrome. Age of onset was between 25 and 30 years.

Histidine-58   Amyloidosis in the Maryland/German kindred, which can be traced back to German immigrants in the 1700s, is associated with a histidine substitution at position 58 of the transthyretin protein. The polyneuropathy that often starts after age 40 was originally designated type II because it starts in the upper extremities with carpal tunnel syndrome. The neuropathy may evolve to a generalized sensory motor neuropathy in all four extremities. Varying degrees of cardiac involvement may occur. Affected individuals often have 15–20 years of severe neuropathic symptoms and gastrointestinal dysfunction.

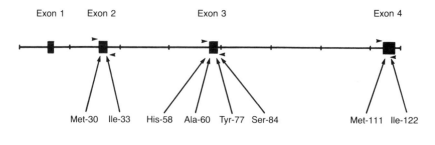

Figure 2.5: Diagram of the prealbumin gene showing four exons (dark boxes) and the characterized mutations associated with hereditary amyloidosis. Arrowheads denote the placement of PCR primers for diagnostic testing.

Alanine-60 Amyloidosis associated with an alanine at position 60 in the transthyretin gene was originally described in a large kindred from West Virginia; hence it was named the Appalachian type. The inheritance of this syndrome was traced back to the late 1700s with Irish, English, and German ancestry. The finding of this type of amyloidosis in other Irish families suggests that the gene was brought to the United States from Ireland. The peripheral neuropathy in this syndrome is similar to the Portuguese FAP I

Table 2.2:   Transthyretin Amyloidoses

Amino Acid Substitution	Position	Clinical Name	Clinical Feature	Kindreds
Methionine	30	FAP I	LLN, bowel, AN, eye	Portugal, Japan, USA, Sweden
Isoleucine	33	Jewish	LLN, eye	Israel
Histidine	58	FAP II/Maryland	CTS, heart	USA/Maryland
Alanine	60	Appalachian	Heart, CTS	USA/West Virginia
Tyrosine	77	Illinois/German	Kidney, bowel, LLN	USA/Illinois
Serine	84	FAP II/Indiana	CTS, heart, eye	USA/Indiana
Asparagine	90	Italian	LLN, eye	USA
Methionine	111	Danish	Heart	Denmark
Cysteine	114	Japanese	LLN, eye	Japan
Isoleucine	122	Senile cardiac	Heart	USA, Scandinavia

LLN = lower-limb neuropathy; AN = autonomic neuropathy; CTS = carpal tunnel syndrome.

but there is a higher incidence of carpal tunnel syndrome, and in some patients the neuropathy is of minor significance. The main features of this syndrome are restrictive cardiomyopathy and gastrointestinal amyloidosis. Age of onset is usually after 50 and in some individuals in their 60s. The disease has a progressive course of 10–15 years with death often after age 60. A few gene carriers have been documented to live past 80 or 90 years of age.

Tyrosine-77   The tyrosine-77 transthyretin variant was originally described in a German family from Illinois. The syndrome starts as a lower-limb neuropathy after age 50. Renal failure has been a major cause of death in the family. Subsequently, a number of other families have been described with this mutation, including one family in France with restrictive cardiomyopathy and two families in the western United States.

Serine-84   The Indiana/Swiss type of amyloidosis is associated with a substitution of serine for isoleucine at position 84 of transthyretin. It was originally reported because of vitreous amyloidosis and, indeed, practically all patients with this mutation develop vitreous deposits of amyloid. The neuropathy typically starts with carpal tunnel syndrome and was therefore designated FAP II to distinguish it from the lower-extremity neuropathy of FAP I. Restrictive cardiomyopathy is the usual cause of death in this family and may occur from age 50 through age 70.

Asparagine-90   A substitution of asparagine for histidine at position 90 of transthyretin has been reported in an Italian woman with amyloidosis. This person had cardiomyopathy and vitreous amyloidosis, but only a mild sensory neuropathy and carpal tunnel syndrome. Onset of symptoms was between age 30 and 40 years with death 5–10 years later.

Methionine-111   This syndrome, which is characterized by restrictive cardiomyopathy, was described in a Danish family in 1962. This is associated with a methionine at position 111 of the transthyretin protein. No peripheral neuropathy has been shown in this syndrome; although amyloid deposits in organs other than the heart have been described post mortem.

Cysteine-114   A Japanese kindred has been recently described with amyloid polyneuropathy presenting shortly after age 30 and resulting in death within 10 years. Vitreous deposits and autonomic neuropathy are features of this syndrome.

Isoleucine-122   A substitution of isoleucine for valine at position 122 in transthyretin has been reported in two separate families with what previously has been termed senile cardiac amyloidosis. In one family, peripheral neuropathy was minor but present, and the patient died at age 74 after more than 10 years of progressive congestive heart failure. This mutation is of interest because both patients reported thus far have been shown to be homozygous for the isoleucine-122 variant gene, and both individuals were black.

In addition to the syndromes described above, there are most likely other transthyretin mutations that are associated with amyloidosis. Also, mutations in transthyretin have been reported that have not been associated with amyloidosis. A serine for glycine substitution at position 6 has been found in a family with hyperthyroxinemia. A threonine for alanine substitution at position 109 has also been found to be associated with hyperthyroxinemia. A substitution of methionine for threonine at position 119 has been discovered which seems to have no association with a pathologic condition.

Apolipoprotein A-I Amyloidosis

One form of FAP that was described in the 1960s in a kindred from Iowa presented as a peripheral neuropathy that was similar to the FAP I of the Portuguese variety. A very strong association with peptic ulcer disease often resulting in gastric perforation led to a separate classification of this syndrome as FAP III. This attention to detail has proven to be justified since it was recently discovered that the amyloid deposits in this kindred contain a degradation product of apolipoprotein A-I. The FAP III syndrome is autosomal dominant. The age of onset is variable, with some individuals being affected in their 20s while others have no symptoms until their 40s or 50s. Renal amyloid deposition is typical for this syndrome, with the deposits being more interstitial than glomerular. For this reason azotemia without the nephrotic syndrome appears to be the rule. The peripheral neuropathy starts in the lower extremities with sensorimotor defect. The association with gastric ulcers is very high, but to this day unexplained.

After the original description in 1969, the kindred was not investigated until 1988 when amyloid material became available for chemical analysis. These fibrils contain an amino-terminal fragment of apolipoprotein A-I with an arginine for glycine substitution at position 26 from the amino terminus. Subsequent evaluation of the family including DNA analysis for patients who died in the 1960s has shown inheritance of the variant apoplipoprotein A-I gene and proven its association with the amyloidosis. Recently, an unrelated family has been discovered with this variant apolipoprotein A-I.

Chemistry of Localized Forms of Hereditary Amyloidosis

Several forms of amyloidosis are considered to be localized in that they involve only one organ system. This includes the amyloid associated with medullary carcinoma of the thyroid, in which the fibrils are composed of a fragment from procalcitonin. The amyloid deposits are limited to the thyroid or the tumor metastases. Cutaneous amyloid has been reported as an autosomal dominant condition, but chemical characterization has not been completed in most cases. Two localized forms of nervous system amyloidosis have been chemically characterized recently: hereditary cerebral hemorrhage with amyloidosis (HCHWA) and Alzheimer's disease (AD). HCHWA was originally described in a large kindred in Iceland in which affected individuals suffered repeated cerebral hemorrhages. Postmortem examinations showed congophilic angiopathy, and chemical analysis of amyloid isolated from leptomeninges showed an amyloid subunit that was a degradation product of cystatin C (γ trace protein). Subsequently, a specific mutation with glutamine instead of leucine at position 58 of the molecule was shown to be associated with the disease. Affected patients have low levels of cerebral spinal fluid cystatin C.

Alzheimer's disease is another form of localized amyloidosis, with deposits limited to the central nervous system and its blood vessels. Amyloid fibrils in cerebral plaques and leptomeningeal blood vessels contain a fragment of a cell membrane protein. This fragment (β amyloid protein) has only 41 amino acid residues of the carboxy-terminal portion of the membrane protein. Recently three mutations in the β protein have been found in patients with autosomal dominant Alzheimer's disease. A family originally described in Holland with hereditary cerebral hemorrhage, similar to the Icelandic kindred, has now been found to have an inherited form of amyloidosis in which the intracranial vessel deposits are composed of the β amyloid protein. A recent report characterized a glutamine substitution for glutamic acid at position 22 of the amyloid subunit protein in two patients with this syndrome.

MOLECULAR MEDICINE

Diagnostics

With the discovery of specific mutations in the transthyretin gene that are associated with amyloidosis, it became possible to test genomic DNA for carrier status. Most of the mutations that have been described in the transthyretin gene have resulted in changes in the restriction enzyme recognition pattern, and therefore allow direct genomic DNA testing (Table 2.3). With the advent of the polymerase chain reaction (PCR) for amplify-

Table 2.3:   Direct DNA Tests

Variant	Mutation	Restriction Enzyme
Transthyretin		
Met-30	G → A	*Nsi* I
Ile-33	T → A	*Bcl* I
His-58	T → A	None*
Ala-60	A → G	*Pvu* II
Tyr-77	C → A	*Ssp* I
Ser-84	T → G	*Alu* I
Asn-90	C → A	*Sph* I
Met-111	C → A	*Dde* I
Cys-114	A → G	None*
Ile-122	G → A	*Mae* III
Apolipoprotein A-I		
Arg-26	G → C	None*

*Use allele-specific PCR.

ing genomic DNA, it is no longer necessary to do standard Southern analysis (Figure 2.5). When the specific mutation in the transthyretin gene is known for a member of a family that is to be tested, the appropriate exon can be amplified from genomic DNA, and then the restriction pattern can be determined using the enzyme specific for the mutation in question (Figure 2.6). When an amyloid transthyretin variant such as histidine-58 does not change the restriction pattern, an allele-specific PCR test can be used in which a mismatch at the 3' end of an amplification primer allows differentiation between the normal and the variant gene. This method, of course, may be applied to all of the other transthyretin variants but is less reliable than the restriction enzyme tests. Both tests can now be accomplished in less than a day, after DNA is isolated, and neither requires radiolabeling. In addition, the PCR test can be applied to tissues that have been aldehyde-fixed for many years, thus allowing retrospective analysis of kindreds.

The FAP III variant apolipoprotein A-I gene does not show a difference in restriction enzyme pattern from the normal. For this reason, an allele-specific PCR test has been developed and appears to be reliable. This has been cross checked with isoelectric focusing of plasma where the variant apolipoprotein A-I (Arg-26) can be distinguished from normal by its +1 charge difference.

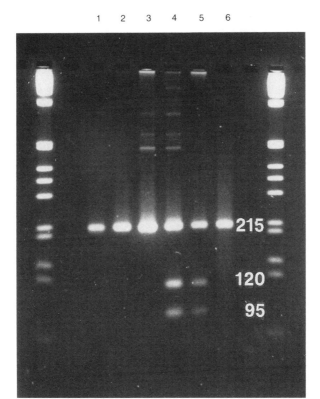

Figure 2.6: Detection of methionine-30 prealbumin gene carriers by PCR amplification of transthyretin exon 2 sequence and digestion with *Nsi* I. Lanes 1, 2, 3, and 6 are DNA from individuals who are not carriers of the methionine-30 gene. Only the normal 215-bp PCR product is observed. Lanes 4 and 5 show *Nsi* I–digested DNA products from individuals who are carriers of the methionine-30 gene. The normal 215-bp product as well as 120-bp and 95-bp fragments, which result from digestion of the variant allele at the new *Nsi* I site, are observed. One kilobase ladder (BRL) molecular weight marker flanks Lanes 1 and 6. DNA fragments were separated by agarose gel electrophoresis, stained by ethidium bromide, and photographed. The entire test can be accomplished in one day. Similar methods are used for the other mutations shown in Figure 2.4.

Therapeutic Options

There is no specific treatment for any form of systemic amyloidosis and this includes the hereditary forms of the disease. Current treatment for systemic amyloidosis is aimed at correcting any metabolic derangement that occurs. This may include treatment of congestive heart failure, bowel dysfunction, renal insufficiency, and infection from neuropathic ulcers. While none of these measures is specific, significant prolongation of life can be expected with proper medical care.

With the advent of direct DNA tests for many of the genes coding for variant proteins associated with amyloidosis, genetic counseling has become more meaningful. As with all autosomal dominant conditions, each child of a gene carrier has a 50% chance of inheriting the variant gene. Since hereditary amyloidosis is usually a late-onset condition, most affected individuals have not benefited from genetic counseling. A few gene carriers have elected not to have children. These are usually individuals who have observed the devastating effects of the disease on their parents. Recently, prenatal testing has become available for the transthyretin-amyloidoses. Both amniocyte DNA and CVS DNA have been used to positively detect prealbumin variant genes in individuals at risk.

Another benefit of DNA testing in hereditary amyloidosis is the prevention of incorrect diagnosis (e.g., immunoglobulin (AL) or reactive AA) in individuals with varying manifestations of systemic amyloidosis. Knowledge of the carrier status can positively affect the medical treatment of these individuals and often prevent inappropriate therapeutic measures.

SELECTED REFERENCES

Andrade C. A peculiar form of peripheral neuropathy. Familial atypical generalized amyloidosis with special involvement of the peripheral nerves. Brain 1952; 75:408.

Andrade C, Araki S, Block WE, et al. Hereditary amyloidosis. Arthritis Rheum 1970;13:902.

Benson MD. Partial amino acid sequence homology between an heredofamilial amyloid protein and human plasma prealbumin. J Clin Invest 1981;67:1035.

Benson MD, Wallace MR. Amyloidosis. In: Scriver CR, Beaudet AL, Sly WS, Valle DV, eds. Metabolic basis of inherited disease, 6th ed. New York: McGraw Hill, 1989:2439–2461.

Blake CCF, Geisow MJ, Oatley SJ. Structure of prealbumin: secondary, tertiary, and quaternary interactions determined by fourier refinement at 1.8A. J Mol Biol 1978;121:339–356.

Chartier-Harlin M-C, Crawford F, Houlden H, et al. Early-onset Alzheimer's disease caused by mutations at codon 717 of the β-amyloid precursor protein gene. Nature 1991;353:844–846.

Costa PP, Figuera AS, Bravo FR. Amyloid fibril protein related to prealbumin in familial amyloidotic polyneuropathy. Proc Natl Acad Sci USA 1978;78:4499.

Dwulet FE, Benson MD. Polymorphism of human plasma thyroxine binding prealbumin. Biochem Biophys Res Commun 1983;114:657.

Falls HF, Jackson JH, Carey JG, Rukavina JG, Block WD. Ocular manifestations of hereditary primary systemic amyloidosis. Arch Ophthalmol 1955;54:660.

Glenner GG. Amyloid deposits and amyloidosis. The fibrilloses. N Engl J Med 1980;302:1283.

Goate A, Chartier-Harlin M-C, Mullan M, et al. Segregation of a missense mutation in the amyloid precursor protein gene with familial Alzheimer's disease. Nature 1991;349:704–706.

Gudmundsson G, Hallgrimsson J, Jonasson TA, Bjarnason O. Hereditary cerebral hemorrhage with amyloidosis. Brain 1972;95:387.

Murrell J, Farlow M, Ghetti B, Benson MD. A mutation in the amyloid precursor protein associated with hereditary Alzheimer's disease. Science 1991;254:97–99.

Nichols WC, Benson MD. Hereditary amyloidosis: detection of variant pre-albumin genes by restriction enzyme analysis of amplified genomic DNA sequences. Clin Genet 1990;37:44–53.

Nichols WC, Dwulet FE, Liepnieks J, Benson MD. Variant apolipoprotein AI as a major constituent of a human hereditary amyloid. Biochem Biophys Res Commun 1988;156:762–768.

Nichols WC, Gregg RE, Brewer HB, Benson MD. A mutation apolipoprotein A-I Iowa type of familial amyloidotic polyneuropathy. Genomics 1990;8:318–323.

Nichols WC, Liepnieks JJ, Benson MD, McKusick VA. Direct sequencing of the gene for Maryland/German familial amyloidotic polyneuropathy type II and genotyping by allele-specific enzymatic amplification. Genomics 1989;5:535–540.

Rukavina JD, Block WD, Jackson CE, Falls HF, Carey JH, Curtis AC. Primary systemic amyloidosis: a review and an experimental, genetic, and clinical study of 29 cases with particular emphasis on the familial form. Medicine 1956;35:239.

Skare JC, Milunsky JM, Milunsky A, Skare IB, Cohen AS, Skinner M. A new transthyretin variant from a patient with familial amyloidotic polyneuropathy has asparagine substituted for histidine at position 90. Clin Genet (in press).

Ueno S, Uemichi T, Yorifuji S, Tarui S. A novel variant of transthyretin (Tyr[114] to Cys) deduced from the nucleotide sequences of gene fragments from familial amyloidotic polyneuropathy in Japanese sibling cases. Biochem Biophys Res Commun 1990;169:143–147.

Van Allen MW, Frohlich JA, Davis JR. Inherited predisposition to generalized amyloidosis. Neurology 1969;19:10.

CHAPTER 3

Duchenne Muscular Dystrophy

Ronald G. Worton
Elizabeth F. Gillard

HISTORIC AND CLINICAL DESCRIPTION

Among the 4,000 or so diseases with a genetic etiology there are a few that stand out because of frequent occurrence and severe clinical phenotype. Duchenne muscular dystrophy (DMD) is one such disease. It is a severe muscle-wasting disorder, resulting in early confinement to a wheelchair and death by the age of 20. Described by Meryon in 1852, Little in 1853, and Duchenne in 1861, it was not until the mid-1950s that diagnostic criteria were developed to allow its clear distinction from other neuromuscular disorders. Becker muscular dystrophy (BMD) resembles DMD but, until recently, it was considered a separate entity because of its later onset, more benign course, and longer survival time. Since DMD and BMD have now been shown to result from lesions in the same X-linked gene they represent opposite ends of the clinical spectrum of the same genetic disease. The history and the clinical features of the disease have been described in numerous reviews, most recently in the book by Emery. Here we describe the clinical features only briefly, with emphasis on new findings relating to clinical progression.

Boys with DMD are phenotypically unremarkable at birth and remain so for the first year or two of life. They present with muscle weakness at age 3–5 when they begin to have difficulty in climbing stairs and in rising from a sitting position on the floor. Confirmation of the diagnosis is usually achieved by measurement of serum creatine kinase (CK) levels, muscle histology, and electromyography (EMG). The serum CK is grossly elevated, especially in the preclinical and early clinical stages of the disease, as a result of the release of the muscle isoform of CK into the serum. Muscle histology characteristic of the disease shows evidence of muscle fiber degeneration and regeneration and includes small fibers, variation in fiber size, invasion by phagocytic cells, and ultimately replacement of the muscle by fat and connective tissue. The EMG is also characteristic, with action potentials reduced in both duration and amplitude, and with enhanced frequency of polyphasic potentials.

While the general features of the natural history of the disease have been well documented for individual cases, there has not, until recently, been extensive systematic recording of the clinical progression in a sizable cohort of affected children. Of some importance, therefore, is the report of Brooke et al. describing a multicenter study designed to measure clinical progression and the effects of supportive therapy. The study included all boys whose symptoms began before age 5; those with the mildest course would have been classified as Becker muscular dystrophy. In the study the age of occurrence of a number of "milestones" was recorded, forming a data base against which an individual patient could be measured. Typically, affected boys lost the ability to rise from a chair and to climb stairs at age 9 (range 7–13 years) and required a wheelchair by age 12 (range 9–16 years). For a period of about 3 years prior to the wheelchair confinement, leg braces were beneficial in maintaining a degree of ambulation.

In the teenage years the major problems were scoliosis and reduction in forced vital capacity (FVC) of the lungs. While 25% of the patients maintained a relatively straight back, progressive scoliosis was common in the others. In most clinics participating in the study, those boys with curvature of $> 35°$ were considered as candidates for corrective surgery (the Luque procedure) to insert a steel rod along the spine. The FVC reduction and the associated risk of pneumonia were greatest in those who scored lowest in performance based on the age at which they reached selected milestones. These "weak" patients, with performance ratings below the 50th percentile, died from respiratory failure and pneumonia with age of death ranging from 13 to 17 years. The stronger boys who scored above the 50th percentile lived a little longer (age of death 14–21 years), and many died of cardiac failure with respiratory function reasonably well preserved. Maintenance of life to age 25 or more is now possible with the aid of a respirator. Despite considerable research effort the disease continues to be devastating and lethal, and it is only through greater understanding of the basic defect that we can expect new approaches to treatment or cure.

MENDELIAN HERITABILITY AND GENE LOCALIZATION

Patterns of Inheritance

Worton and Thompson have reviewed the inheritance of Duchenne and Becker muscular dystrophy. The incidence of DMD is approximately 1 in

3,300 male births with little ethnic variation, and the calculated mutation rate of 10^{-4} is an order of magnitude higher than for most other genetic diseases. Genetic counseling is usually based on the assumption of equal mutation rates in males and females, an assumption that leads to the prediction that one-third of affected males should result from a new mutation while two-thirds should result from inheritance of the defective gene from a carrier mother. Segregation analyses in several large studies are generally consistent with this expectation. BMD accounts for 11–17% of mutations at the locus, giving an incidence of 3–5 × 10^{-5}. Since approximately 70% of males with BMD are capable of transmitting the gene to a carrier daughter, a higher proportion of BMD cases are inherited and fewer (approximately 10%) are new mutants. The mutation rate for BMD is only 3–5% of that for DMD.

Although DMD and BMD are expressed primarily in boys, random X-inactivation in females leads to mild clinical manifestation in about 8% of carriers. An extreme example of this is found in monozygotic female twin carriers who are discordant for the DMD phenotype. When it has been examined, the discordancy has appeared to be the result of preferential inactivation of different X chromosomes in each member of the twin pair. The disease may also be expressed in Turner syndrome females who have a single X chromosome and in females with X-autosome translocations that disrupt the *DMD* gene. In the latter, the translocation not only disrupts the *DMD* gene on one X chromosome but it also causes the nonrandom inactivation of the normal allele on the other X chromosome, resulting in the expression of the disease phenotype.

Gene Localization

Mapping of the *DMD* gene on the X chromosome was the first step toward cloning of the gene and identification of its protein product, *dystrophin*. Three lines of evidence mapping the *DMD* gene to band Xp21 in the middle of the short arm of the X chromosome have been described in some detail elsewhere and will be reviewed here only briefly. The first evidence came from the translocation females described above. Between 1979 and 1983 six such females were identified and in each case the exchange point in the X chromosome was found to be in band Xp21, suggesting that a gene involved in muscular dystrophy might be located at this site. Subsequent high-resolution banding analysis of these and other translocations revealed that the exchange points were not precisely the same in all affected females, and suggested a target for disruption extending from Xp21.1 to Xp21.3, a region of perhaps 3–4 million base pairs (bp). This in turn suggested the possibility of a very large gene, a speculation that turned out to be correct.

The second line of evidence came from family studies with DNA probes that detect restriction fragment length polymorphism (RFLP) on the X chromosome. The first two linked markers for the *DMD* gene were revealed by probes RC8 and L1.28. In Duchenne families both markers segregated with the *DMD* gene but displayed a recombination frequency of about 20%, mapping the markers 20 cM (one cM or centimorgan is the genetic distance corresponding to a recombination frequency of 1%) from the *DMD* gene. RC8 mapped to the distal third and L1.28 to the proximal third of the short arm of the X chromosome, and the two markers mapped about 40 cM apart, suggesting that they must flank the *DMD* gene. This mapped the gene to the middle of the short arm, consistent with the translocation data. Since these initial reports in 1983, many additional linked markers have been identified, and in Becker families the markers all gave segregation results essentially similar to that for Duchenne families, providing the first indication that the *DMD* and the *BMD* genes are closely linked or identical.

The third line of evidence came from a few boys with Duchenne muscular dystrophy in combination with one or more additional X-linked diseases. One of these, a boy who has become well known as "BB," was shown to have DMD, retinitis pigmentosa (RP), chronic granulomatous disease (CYBB), and the McLeod red cell phenotype (XK), all due to a small but cytogenetically visible deletion of part of band Xp21. Other boys with similar phenotypes, or with DMD plus glycerol kinase (GK) deficiency or GK deficiency plus congenital adrenal hypoplasia (AHC) have since been studied and shown to have similar contiguous gene deletions. Several papers review the extent of the various deletions and demonstrate a probable gene order of *AHC, GK, DMD, XK, CYBB, RP* in a telomere-to-centromere direction. Mapping of this block of contiguous genes to Xp21 provided further evidence for the location of the *DMD* gene, and set the stage for the cloning of the gene.

MOLECULAR GENETICS

Cloning of the *DMD* Gene

Cloning of a disease gene from knowledge of its location in the human genome is a process that has been termed *reverse genetics* or *positional cloning* to distinguish it from the more established procedure of gene cloning from knowledge of the RNA or protein product. Genomic clones from the *DMD* locus at Xp21 were isolated by two different strategies.

The PERT87 clone was isolated by Kunkel and colleagues using a competitive hybridization procedure (phenol-enhanced reassociation technique—PERT) to enrich for DNA fragments from the region of the BB deletion. The PERT87 clone was one of a few that mapped into the BB deletion and it was the only one that failed to hybridize with DNA from a subset of boys with DMD. The assumption, which turned out to be correct, was that these boys had the disease as a consequence of a submicroscopic deletion that removed at least a portion of the *DMD* gene, including the PERT87 sequence. The cloned region was expanded by walking from PERT87 through the isolation of a series of overlapping clones, and the 220-kb cloned region (locus *DXS164*) was scanned for the presence of exon sequences that might encode a portion of a protein. Since exons are more conserved during evolution than are introns, subclones were tested for their ability to hybridize to DNA from multiple species in a "zoo" blot. One small region with this property was found to hybridize with a 14-kb mRNA from skeletal muscle, and was used as a probe to isolate a fetal muscle cDNA clone. The cDNA hybridized to eight genomic fragments, each containing one or more exons of a gene spread over the 220-kb *DXS164* locus. Since the cDNA clone represented only one-fourteenth of the coding sequence, the complete gene was estimated to be 2,000–3,000 kb in size, a factor of 10 larger than any known gene. The remainder of the transcribed sequence was isolated in a series of overlapping cDNA clones that were sequenced and found to encode a protein of 3,685 amino acids, with a predicted molecular weight (MW) of 426 kD.

In our own laboratory a genomic clone XJ1 was isolated from the junction of a t(X;21) translocation carried by a Belgian girl with muscular dystrophy. Her translocation had an exchange point in chromosome 21 in a block of tandemly repeated genes encoding ribosomal RNA. Ribosomal gene probes were used to identify a clone containing a portion of ribosomal gene at its telomeric end and a segment of the X chromosome at its centromeric end. The XJ1 clone also failed to hybridize with the DNA from a subset of affected boys, many of them the same ones that failed to hybridize with the PERT87 probe, providing further support for the proximity of XJ1 to the *DMD* locus. Chromosome walking from XJ1 yielded the 140-kb *DXS206* locus, a subclone of which was used as a probe to isolate a muscle cDNA clone. This clone hybridized with 13 genomic fragments containing 15 exons of a large gene. Eight exons mapped into the *DXS164* locus, two mapped into the *DXS206* locus, and the remaining seven mapped on the centromeric side of these loci. These turned out to be the first 15 exons of the same large gene isolated by the Kunkel group.

Because of the large size of the *DMD* mRNA, the cDNA was isolated as a series of overlapping clones rather than as a single clone. The clones most often used for patient analysis are those isolated in Kunkel's laboratory. These clones, designated 1-2a, 2b-3, 4-5a, 5b-7, 8, and 9-14, are shown in Figure 3.1 as bars along a schematic of the exons. Other probes in use for patient analysis are also shown in Figure 3.1 and were isolated in our own laboratory (D38, 10-69, D43, and 46-6) or Davies' laboratory (Cf and Ca series). A list of the exons and the exon-containing *Hin*dIII fragments detected by the complete cDNA is provided in Table 3.1.

Figure 3.2 presents a schematic of the *DMD* gene oriented in a centromere-to-telomere direction. The location of the PERT87 and the XJ1 sequences are indicated, as are other cloned regions that serve as landmarks in this remarkably large gene. The approximate size of the gene and the location of the various cloned parts were determined by pulsed field gel analysis of large fragments. Only the *Sfi* I fragments are shown, but it does illustrate that the gene is enormous, covering more than 2,300 kb, or approximately 1.5% of the X chromosome. The gene is 100-fold larger than a typical gene, is half the size of the entire *E coli* genome, and is larger than any of the chromosomes of baker's yeast. Since the gene is about the same size in rodents and birds, the size has been maintained through evolution. The question is—why? What were the driving forces that created such a large gene, and how has it maintained its size through evolution?

Dystrophin, the Product of the *DMD* Gene

To identify the product of the *DMD* gene, the general approach has been to examine the amino acid sequence in order to identify homologies with

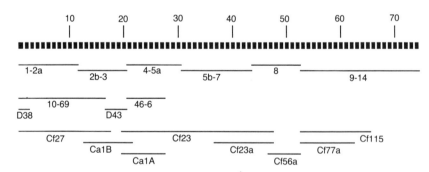

Figure 3.1: A schematic indicating the position of cDNA clones relative to the exons of the *DMD* gene. Each exon detected by the *DMD* cDNA is shown by a box (not to scale). Below these boxes are lines indicating the exons detected by each cDNA clone. Clones from Kunkel's laboratory are labeled 1-2a, 2b-3, 4-5a, 5b-7, 8, and 9-14, indicating their position along the 13.9-kb message. Clones from our laboratory are labeled D38, 10-69, D43, and 46-6. Clones from Davies' laboratory are labeled with prefix Cf (fetal) or Ca (adult) depending on the source of the muscle cDNA library. Exon numbers are as given in Table 3.1.

Table 3.1: Order, *Hind*III Fragment Size, and 3' Border Type of Exons

Exon Number	HindIII Frag. Size	3' Border Type	Exon Number	HindIII Frag. Size	3' Border Type
1	3.20	1	38	6.00	—
2	3.25	3	39	"	—
3	4.20	3	40	6.20	—
4	8.50	3	41	"	3
5	3.10	3	42	4.20	3
6	8.00	2	43	11.00	2
7	4.60	1	44	4.10	3
8	7.50	3	45	0.50	2
9	"	3	46	1.50	3
10	10.50	3	47	10.00	3
11	"	2	48	1.25 + 3.8	3
12	3.90	3	49	1.60	3
13	6.60	3	50	3.70	1
14	2.70	3	51	3.10	3
15	"	3	52	7.00	1
16	6.00	3	53	7.8 + 1.0	3
17	1.70	2	54	8.30	2
18	12.00	3	55	2.30	3
19	3.00	1	56	1.00	—
20	7.30	3	57	8.80	—
21	11.00	1	58	6.00	—
22	20.00	3	59		3
23	"	3	60	3.50	3
24	"	3	61	6.60	—
25	"	3	(62)	2.80	—
26	5.20	—	(63)	12.00	—
27	"	—	(64)	2.40	—
28	12.00	—	(65)	2.55	—
29	4.70	—	(66)	1.45	—
30	18.00	—	(67)	1.50	—
31	"	—	(68)	6.80	—
32	"	—	(69)	2.10	—
33	"	3	(70)	1.90	—
34	1.80	3	(71)	10.00	—
35	0.45	3	(72)	1.80	—
36	1.30	3	(73)	5.90	—
37	1.50	3	(74)	7.80	—

The order of *Hind*III fragments, their assigned exon numbers, and their 3' border types are as described by Koenig et al., 1989 with the following exceptions:

1. The 3' border types for exon's 22–25 were defined by Karen Bebchuk in our laboratory (unpublished data).
2. The order of exons 28 and 29 is reversed based on deletion data from our laboratory.
3. The order of exons 61–65 differs from that of Koenig et al., 1989 and is based on data from a deletion patient and a translocation patient from our laboratory, the order of 62 and 63 being uncertain.
4. The numbering and order of exons 66–75 is uncertain and is based on a consensus reached at a 1989 Banbury conference on dystrophin.

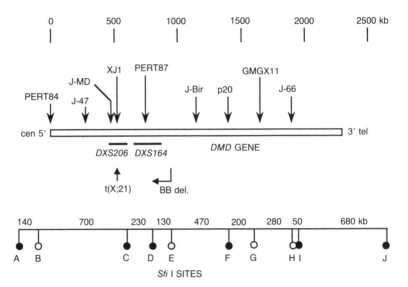

Figure 3.2: A schematic of the 2.3 Mb *DMD* gene. The gene is oriented from 5′ → 3′ in a centromere (cen) to telomere (tel) direction. The 3′ breakpoint of the deletion in "BB" is shown, as is the t(X;21) translocation exchange point. The locations of the cloned genomic markers that serve as landmarks are indicated by arrows. The PERT87 and XJ1 clones are described in the text. The regions *DXS164* and *DXS206* were isolated by chromosome walking from PERT87 and XJ1, respectively. Other important landmarks include the p20 clone and the GMGX11 clone, isolated from genomic DNA libraries enriched for all or part of the X chromosome. To isolate the "jump" clones, J-47, J-MD, J-Bir, and J-66, a DXS164 probe was used to identify and clone a segment containing the deletion junction from DNA of patients with deletions extending from the *DXS164* locus to the region of the jump clone. The PERT84 clone was isolated in the same manner as the PERT87 clone, and was later recognized to map at the 5′ end of the gene. Below the gene is the map of *Sfi* I sites as determined by pulsed field gel electrophoresis (adapted from den Dunnen et al.). The open circles represent partially digestible *Sfi* I sites, whereas the solid circles represent fully digestible *Sfi* I sites.

other known proteins and to generate antibodies against epitopes on the protein for use in both Western blot analysis and immunocytochemistry. Polyclonal antibodies have been raised in animals immunized with either synthetic peptides conjugated to larger molecules, or fusion proteins synthesized by bacteria that had been transfected with a gene containing a portion of the *DMD* gene joined to the 5′ end of a highly expressed bacterial gene. Several monoclonal antibodies directed against portions of the dystrophin molecule have also been generated.

Sequencing of the *DMD* cDNA predicted a protein of 3,685 amino acids with four distinct domains, as depicted in Figure 3.3A. The protein was named dystrophin to relate it to its role in preventing muscular dystrophy. The N-terminal domain (amino acids 14 to 240) shows sequence homology to the actin-binding domain of α-actinin. A large second domain (amino acids 278 to 3,080) has a 109–amino acid repeat pattern similar to the repeats found in α-actinin and spectrin and thought to form a triple helical rod-like segment. The third domain (amino acids, 3,080 to 3,360) is cysteine-rich and also bears some homology to α-actinin. Recently, an autosomal transcript with homology to the C-terminal domain of dystrophin has been described, as discussed below. The features of dystrophin are highly suggestive of a cytoskeletal protein, perhaps with a structural role in the cell.

Antidystrophin antibodies identify a protein of about 400,000 Daltons on Western blots of human and mouse skeletal muscle (Figure 3.3B). Dystrophin is also present in cardiac muscle, with lesser amounts in smooth muscle and brain. The protein is missing from muscle of boys with Duchenne muscular dystrophy, and from *mdx* mice with a mutation in the equivalent mouse gene (Figures 3.3B and 3.3D). Biochemical fractionation suggested localization of dystrophin in the triads of skeletal muscle since immunoreactive dystrophin appeared to copurify with other proteins of the triad junction. This, in turn, suggested a possible role for dystrophin in excitation-contraction coupling and Ca^{2+} homeostasis. However, subsequent studies with immunofluorescence and immunocytochemistry, using both light microscopy and electron microscopy, have revealed that the major site of dystrophin is at the inner surface of the sacrolemmal membrane (Figures 3.3C and 3.3E), suggesting a possible role for dystrophin in maintenance of the structural integrity of the membrane. This is an attractive hypothesis because contraction-induced tearing of a weakened membrane in boys without dystrophin might well be the cause of segmental necrosis of muscle leading to the repeated cycles of degeneration and regeneration characteristic of the disease.

One of the most important domains of the dystrophin molecule is the C-terminal domain. This became clear with the cloning of the chicken dystrophin gene and the finding that the greatest sequence conservation is in the C-terminal domain of the protein and in the 3′ untranslated portion of the gene. This suggests that the C-terminal end of the protein has an important biological function, and that the 3′ untranslated region must play some, as yet undetermined, function in the expression of the gene.

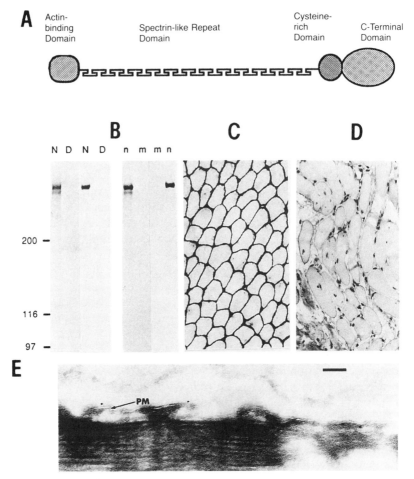

Figure 3.3: Structure and localization of dystrophin. (A) A schematic of the dystrophin molecule showing the four domains described in the text. (B) Western blots of muscle from normal human (N), Duchenne (D), normal mouse (n), and *mdx* mouse (m). Lanes 1, 2, 5, and 6 were stained with the antibody 9219, directed against amino acids encoded by exons 4–16 near the amino terminus of dystrophin; lanes 3, 4, 7, and 8 were stained with antibody 1461, directed against the last 17 amino acids at the carboxy terminus. (C) Section of normal muscle stained with antidystrophin antibody 9219 and counterstained with hematoxylin. Note the intense staining at the periphery of each myofiber. (D) Section of Duchenne muscle stained with antidystrophin antibody 9219 and counterstained with hematoxylin. Note absence of dystrophin staining at the periphery of the myofiber. (E) Electron micrograph of a human muscle fiber (gastrocnemius) labeled with antidystrophin antibody 1460, directed against the last 17 amino acids of dystrophin. Immunoreactive dystrophin is labeled with gold particles that appear along the plasma membrane (PM). Bar = 100 nm. *Figures 3.3B, C, D courtesy of Dennis Bulman; Figure 3.3E courtesy of Dr. Michael Cullen.*

Further significance of the C-terminal domain is suggested by the observation that cDNA fragments encoding this part of the protein detect a closely related gene that exhibits nucleic acid and predicted amino acid sequence similarities with the *DMD* gene. This dystrophin-related gene, reported by Love et al., encodes a transcript of 13 kb, maps to human chromosome 6, and is a logical candidate for the defective gene in one of the autosomally inherited dystrophies.

New insight into the potential function of dystrophin has come from its purification on a wheat germ agglutinin (WGA) column, complexed with five integral membrane glycoproteins. Campbell has suggested that the localization of dystrophin at the cytoplasmic face of the sarcolemma results from its tight association with this glycoprotein complex and that the complex may serve as a link between the membrane and the internal cytoskeleton. Indeed, Ervasti and Campbell have demonstrated that dystrophin deficiency is accompanied by loss of the glycoprotein complex at the membrane and that one protein of this complex, the 156-kD "dystroglycan," binds to laminin on the outside of the muscle fiber. These observations would support such a model.

Tissue Specificity and Regulation of the *DMD* Gene

The availability of *DMD* cDNA clones and antidystrophin antibodies has prompted a number of studies aimed at defining the tissue specificity and developmental pattern of *DMD* gene expression.

Tissue and Developmental Specificity   Initially, Northern blot analysis revealed a low-abundance, 14-kb mRNA transcript in fetal and adult skeletal muscle. More sensitive techniques such as RNase protection and polymerase chain reaction have clearly indicated that the *DMD* transcript is most abundant in cardiac and skeletal muscle, with reduced amounts in smooth muscle and brain, a distribution that is in good agreement with the clinical spectrum of tissue involvement in the disease. In terms of developmental expression, dystrophin transcripts are present in both fetal and adult tissues, and in myogenic cell cultures transcription is initiated as myoblasts begin to differentiate into multinucleated myotubes. Only a small amount of "illegitimate transcription" is seen in other tissues or in cultured lymphoblasts. The *DMD* gene is therefore regulated in a tissue- and stage-specific manner.

At the protein level, Western blot analysis with antidystrophin antibodies has revealed dystrophin in extracts of adult and fetal skeletal, cardiac and smooth muscle, with an apparently lower molecular weight isoform in smooth muscle tissue. Lower levels have been detected in brain and neuronal cell cultures. In adult skeletal and cardiac muscle, dystrophin is localized at the muscle cell membrane without preferential distribution to any particular fiber type, and in myogenic cultures immunostaining of dystrophin is observed at the membrane of myotubes following initiation of differentiation. In contrast, dystrophin staining has been reported in fetal muscle at the ends of the myotubes between the distal nuclei and the myotendonous junction.

Some of the tissue and developmental differences might be the result of isoforms, since for muscle proteins, the generation of isoforms seems to be the rule rather than the exception. The first evidence for this was the finding in rats of a brain-specific promoter resulting in transcription of a different first exon in brain compared with muscle, a result that was later confirmed for the human gene. Another common mechanism for generating isoforms is alternative splicing, resulting in a spectrum of mRNA molecules with different combinations of exons spliced into the message. The finding of alternative splicing in the *DMD* gene confirmed the existence of dystrophin isoforms with at least two different forms in skeletal muscle, three in brain, and two in smooth muscle. The alternative splicing all took place near the 3′ end of the gene, resulting in different amino acid sequences in the carboxy-terminal domain of the protein. Two of the alternative splice patterns resulted in a reading frame shift, predicted to give rise to a truncated protein or a protein ending with a completely different sequence of amino acids.

Regulation of the *DMD* Gene   Gene regulation in myogenic differentiation is controlled by a small number of tissue-specific regulators that interact with conserved enhancer elements located in the promoter regions of muscle-specific genes to coordinately activate their transcription. One such element in muscle genes is the MEF-1 (myocyte-specific enhancer-binding nuclear factor 1) consensus. MEF-1 sites bind MyoD and have been shown to be necessary for full muscle-specific gene expression. Both single and paired MEF-1 sites are observed within the *DMD* gene promoter. Other elements include the sequence $CC(AT-rich)_6GG(CArG$ box), identified in cardiac and skeletal α-actin genes and subsequently in myosin light chain, cardiac troponin T, and muscle creatine kinase genes; and the MCAT (muscle-CAT) sequence initially identified in the chicken cardiac troponin T gene and found in several other muscle-specific genes.

Figure 3.4, adapted from Klamut et al., presents the sequence of 850 bp of the *DMD* gene muscle-specific promoter upstream of the transcription initiation site, and the sequence of 537 bp that includes exon 1 and a portion of intron 1. Exon 1 (underlined in Figure 3.4) is 275 bp in size, of which 244 bp corresponds to the 5′ untranslated region, and 31 bp,

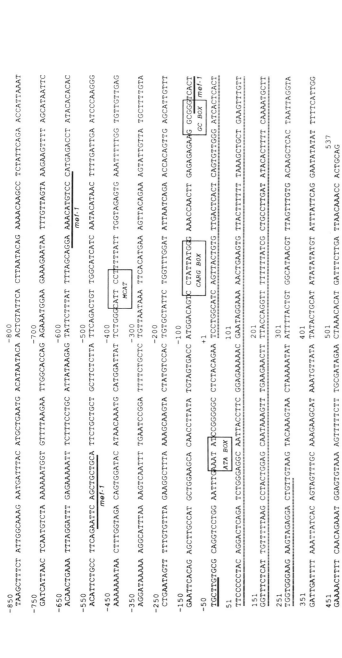

Figure 3.4: Sequence of the *DMD* muscle-specific promoter and the first exon of the *DMD* gene. Shown is 850 bp of upstream sequence and 537 bp of the *DMD* gene sequence. The regulatory elements referred to in the text (ATA box, GC box, CArG box, and MCAT consensus sequences) are boxed. Positions of consensus sequence homologies for MEF-1 binding are underlined solid. The first exon is underlined. The putative start codon for initiation of protein synthesis (the ATG encoding the methionine at the N terminus) is at position 245. *Adapted from Klamut HJ, Gangopadhyay SB, Worton RG, Ray PN. Molecular and functional analysis of the muscle-specific promoter region of the Duchenne muscular dystrophy gene. Mol Cell Biol 1990;10:193–205.*

starting at position 245, includes the translation initiation site and encodes the first 10 amino acids of the protein. The first intron, which begins at position 276, is estimated to be at least 100 kb in size.

The muscle-specific *DMD* gene promoter has been shown to contain sequence elements homologous to those regulatory elements described above. As indicated in Figure 3.4, three MEF-1 sites are located at positions −58, −535, and −583 bp, a single CArG box at position −91 bp, and an MCAT consensus sequence at position −394 bp. Also indicated are an AAATAT sequence (TATA box) at −24 bp and a GC box at −61 bp, which are *cis*-acting elements common to many tissue-specific gene promoters. Confirmation that this promoter functions in a muscle-specific fashion has been obtained by joining it to a "reporter gene." Enhanced transcription of the reporter gene is observed upon differentiation of myoblasts transfected with the *DMD* promoter–reporter gene construct.

The promoter described in Figure 3.4 is active in muscle and the first exon is the one found in muscle mRNA. As described above, the brain transcript is initiated from a different promoter located upstream from the muscle-specific promoter. The brain message therefore has a different first exon and encodes different amino acids at the beginning of the protein. Since the first exon of the brain message is spliced to the same second exon as in the muscle message, the remainder of the brain protein is expected to be essentially identical to muscle dystrophin, although alternate splicing is known to occur at the 3′ end of the gene. A third promoter in the gene is suggested by the recent finding of a short 6.5-kb transcript in nonmuscle tissues by Bar et al.

Mutations in the *DMD* Gene: Deletions, Duplications and Translocations

Deletion of one or more exons of the *DMD* gene accounts for over 60% of mutations associated with DMD or BMD. Duplications account for another 6% of mutations in the gene. The first deletions and duplications described were detected with the PERT87 and the XJ series of genomic probes, and they occurred in about 10–12% of affected boys. In the past three years, however, most deletions and duplications have been detected with a series of cDNA probes containing sequences from the 74 or more exons of the gene. With these probes the frequency of deletion rose to over 60%.

A broad spectrum of deletions has now been described in DMD and BMD patients. The largest deletions, several thousand kilobases in size, often remove neighboring genes and occur in association with a contiguous gene syndrome. Smaller deletions usually remove from one to a few exons and their distribution in the gene is not random. Two deletion-rich regions are apparent, as shown schematically in Figure 3.5. The first is a broad region with large deletions, many overlapping the first 20 exons detected by cDNA clones 1-2a and 2b-3. Many of these deletions were detected by the original PERT87 and XJ probes. The second deletion-rich region occurs near the middle of the gene and encompasses genomic clones P20 and GMGX11 and exons 44–53. Between these two "hot spots" is a region of the gene relatively free of deletions, and in the last quarter of the gene deletions are almost never detected. Deletions in the region between the hot spots have recently been described in individuals

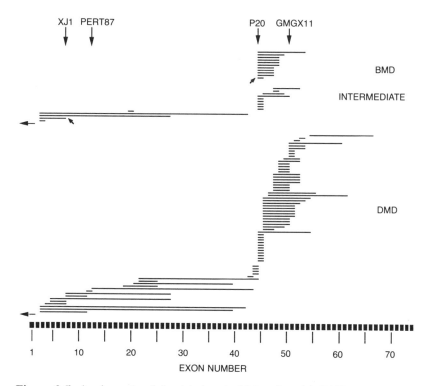

Figure 3.5: A schematic of the deletions in 100 males with DMD, BMD, or an intermediate phenotype. Each line represents the extent of one deletion, plotted along the map of the exons numbered as in Table 3.1. The arrows at the left indicate deletions extending 5′ from exon 1, into the promoter region. The angled arrows indicate two patients referred to in the text, a BMD patient with a deletion of exon 45, and an intermediate patient with a deletion of exons 3–7. The positions of genomic clones PERT87, XJ1, P20, and GMGX11 are also indicated.

with muscle cramps but no muscle weakness. Thus, regions of the *DMD* gene not commonly deleted in DMD/BMD patients may instead be associated with distinct phenotypes, and the total spectrum of deletions may be more uniformly distributed across the gene than indicated by studies of DMD/BMD alone. Duplication of one or more exons constitutes an additional 6% of mutations; these duplicated exons are heterogeneous with respect to both size and location.

The remaining 25–35% of mutations have yet to be defined. They may include point mutations that result in a nonsense (stop) codon, mutations that create or destroy splice sites to cause alternative splicing, or mutations in the muscle-specific promoter to down-regulate its expression. Point mutation to create a stop codon, presumably resulting in premature termination of dystrophin synthesis, has been demonstrated in the *DMD* gene of *mdx* mice, the murine analog of DMD associated with complete absence of dystrophin in skeletal muscle. Recently D. Bulman, in our laboratory, has described a point mutation resulting in a stop codon in exon 26 of the *DMD* gene transcript in a boy who had displayed truncated dystrophin by Western analysis.

We have already discussed females with Duchenne or Becker muscular dystrophy caused by X;autosome translocations that were thought to disrupt the *DMD* gene at Xp21. Several translocation exchange points have now been analyzed and all have been shown to lie within the *DMD* gene as anticipated.

Once it became possible to analyze individual mutations, the potential existed to correlate the size or location of a deletion or duplication to the phenotype of the patient. Mental capacity is one highly variable phenotype in DMD. While many affected boys are of average or above average intelligence, the overall mean IQ for DMD is about one standard deviation below the mean. On the other hand, BMD patients are rarely mentally retarded. So far there is no clear correlation between impairment of mental function and the location or size of deletions or duplications, and the relationship between DMD and mental retardation remains one of the unsolved puzzles associated with this disorder.

Molecular and Clinical Heterogeneity—The Frame-Shift Hypothesis

Although once thought to be distinct diseases, DMD and BMD are now recognized to represent opposite ends of the severity spectrum of the same disease. One of the more interesting questions to be addressed successfully by molecular analysis is the mechanism by which mutations in one gene give rise to the full spectrum of phenotypes ranging from severe DMD to mild BMD.

Even from the earliest studies with the PERT87 and XJ genomic probes it was apparent that deletions in the *DMD* gene were highly heterogeneous with respect to both size and location. BMD deletions were not confined to a specific region of the gene, indicating that differences in phenotype are not due to deletion of discrete domains with differing functional significance. Furthermore, phenotypic severity is not simply a function of deletion size, since some deletions associated with BMD are larger than, and completely encompass, deletions associated with DMD (Figure 3.5). Indeed, one deletion has been found to remove 46% of the DMD-coding sequence in a 65-year-old man with BMD mild enough to allow him to still be driving his automobile.

These observations required explanation and led to the idea that severity of the phenotype might be a direct consequence of the effect of the mutation on the translational reading frame of the mRNA. According to the model, outlined schematically in Figure 3.6, mild disease (BMD) might result from a deletion of one or more exons containing an integral

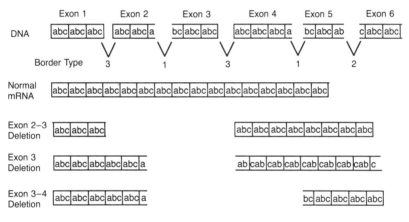

Figure 3.6: Schematic of the frame-shift hypothesis for a hypothetical gene. Each triplet codon is indicated by a box and the letters "a," "b," and "c" are used to represent the first, second and third position in the codons of the normal mRNA. The exon-intron borders that occur after the first, second, or third nucleotide of a codon are indicated as 1, 2, or 3, respectively. The normal mRNA, corresponding to translation of the six hypothetical exons, is indicated below the DNA. The exon 2–3 deletion juxtaposes borders of type 3 and thus maintains the translational reading frame. The exon 3 deletion juxtaposes a type 1 border with a type 3 border, causing the translational reading frame to shift. In the final example, an exon 3–4 deletion juxtaposes borders of type 1, maintaining the translational reading frame. In the last example, the new codon created at the deletion junction is made up of parts of two different codons and could therefore be a stop codon, resulting in premature termination of the protein.

number of codons, since this would maintain the translational reading frame in the mRNA and produce a dystrophin molecule with an internal deletion but intact ends. Severe disease (DMD) might result from a deletion of a set of exons that contain a nonintegral number of codons, resulting in an altered reading frame in the mRNA sequence following the deletion. In this situation translation of the message on the 3' side of the deletion would insert incorrect amino acids into the growing polypeptide until a stop codon was reached, causing premature termination of the protein. The altered dystrophin molecule might be unstable, might fail to localize at the membrane and/or might fail to bind the associated glycoproteins, the net result being loss of dystrophin function. Intermediate patients could fall into either category or alternatively might be found to have in-frame deletions of functionally important domains.

The frame-shift model would be expected to apply equally well to duplications, with duplication of an integral number of codons causing a mild phenotype and duplication of a nonintegral number of codons causing a frame-shift mutation and a severe phenotype. Deletions or duplications comprising any integral multiple of three base pairs, but not beginning and ending at the start of a codon, will in general, maintain the translational reading frame, but occasionally juxtaposition of flanking exons may create a nonsense codon and result in premature termination.

Testing of the model required the determination of exon sequence and the precise location of exon/intron borders in order to predict the effect of deletions or duplications on the translational reading frame of the mRNA. Exon/intron borders have now been defined for over half of the *DMD* gene (Table 3.1). A convenient way of examining deletions and duplications is to compare the "border type" at the 3' end of the exon preceding the deletion (or duplication) and at the 5' end of the exon following the deletion (or duplication). The concept is explained in Figure 3.6. Briefly, an exon border type can be designated "1," "2," or "3" depending on whether it occurs after the first, second, or third nucleotide of a codon. If a deletion results in the splicing of the exon immediately preceding the deletion to the exon immediately following the deletion, *and* if the splicing joins two borders of the same type, then the result is an mRNA that retains the correct reading frame. Conversely, a splicing of two exons with different border types will give rise to a message with an altered reading frame. The known 3' exon border types are given in Table 3.1.

Several groups have evaluated their patients in relation to the frame-shift model and found the hypothesis to hold true for the majority of deletions and duplications. The deletion results are summarized in Table 3.2, and indicate that most of the time the reading frame status will provide a valid prognostic indicator for the severity of the disease. Exceptions

Table 3.2: Summary of Published Frame-Shift Data

Phenotype	DMD		Intermediate		Becker		Reference
Category	Frame-Shift	In-Frame	Frame-Shift	In-Frame	Frame-Shift	In-Frame	
Region Analyzed							
Exons 1–10	11 (2*)	0	8 (5*)	2	6 (6*)	0	Malhotra et al., 1988
Exons 1–21	5	0	2 (2*)	1	0	0	Baumbach et al., 1989
Exons 44–51	20	0	2	0	1 (1†)	1	Baumbach et al., 1989
Entire gene‡	178 (5*,6†)	10	6 (3*)	8	4 (4*)	53	Koenig et al., 1989
Exons 44–51	21 (3†)	0	7 (4†)	1	1 (1†)	8	Gillard et al., 1989
Total	235	10	25	12	12	62	
% of phenotype	96%	4%	68%	32%	16%	84%	
% of category§	87%	12%	9%	14%	4%	74%	

*Number of patients that had exon 3–7 deletions.
†Number of patients that had exon 45 deletions.
‡Excluding regions with an end point(s) having an unknown exon border (10 patients excluded).
§Proportion of patients in the frame-shift or in-frame category.

to the frame-shift hypothesis do exist and need further explanation. One of these is seen in Figure 3.5 as a deletion of exon 45 in 11 DMD patients, 5 intermediate-phenotype patients and one BMD patient (arrow in Figure 3.5). This deletion is expected to cause a frame-shift in the message (joins a type 3 border to a type 2 border, Table 3.1) and the BMD patient with this deletion requires explanation.

Another notable exception to the frame-shift model is a group of patients with deletions for exons 3–7. This deletion, assuming that exon 2 (3' border type 3) is spliced to exon 8 (5' border type 1), is predicted to shift the translational reading frame to cause a severe phenotype; yet patients with this deletion include several with BMD, a few with DMD and some with an intermediate phenotype. Only one of these (in the intermediate class) is among the 100 patients plotted in Figure 3.5 (arrow), the others coming from a large international study. Explanations proposed to account for the mild phenotype in the BMD patients included (1) alternative splicing, e.g., exon 2 to exon 9 or exon 1 to exon 8, both of which maintain the reading frame; (2) initiation or re-initiation of protein synthesis from an in-frame putative translational start site in exon 8; and (3) transcription initiation from an unidentified promoter in intron 7, with protein synthesis initiated in exon 8. A recent study of *DMD* gene transcripts by Chelly et al. has detected alternative splicing in two BMD patients with deletion of exons 3–7. While this was thought to explain the mild phenotype, a similar study by Gangopadhyay in our laboratory has failed to reveal alternatively spliced transcripts in muscle RNA from seven similar patients, suggesting that other mechanisms may also be operating.

Molecular Concepts: Mechanisms of Deletion, Duplication, and Translocation

Deletions are known to occur in many genes and in many genetic diseases; translocations are known to occur in neoplastic cells, where they result in the activation of cellular oncogenes. In some cases it has been possible to determine the molecular details of the rearrangement, providing insight into the mechanism of its occurrence. Examples include recombination between Alu repeats to cause deletion in the LDL receptor gene, and aberrant immunoglobulin gene rearrangement resulting in translocations that activate cellular oncogenes. In the case of the *DMD* gene, the enormous size of the gene, the X-linked form of inheritance, and the frequent occurrence of new mutations combine to provide a unique system for the study of mutational mechanisms.

A priori, the chromosomal rearrangements characteristic of the *DMD* gene may occur in either males or females, in germ cells during meiosis, or in diploid cells during the mitotic cycle, and from either homologous or nonhomologous exchange. These parameters are all amenable to study in DMD families.

As diagrammed in Figure 3.7, deletion mechanisms that may be envisaged include (1) intrachromatid exchange, (2) unequal chromosome exchange (non–sister chromatid exchange for the X chromosome limited to females), or (3) unequal sister chromatid exchange. Any of these could be facilitated by homologous exchange between interspersed repeats (the 5′ and 3′ elements at the site of exchange are homologous) or by nonhomologous exchange, perhaps between specific domains in the chromatin structure (the 5′ and 3′ elements at the site of exchange are nonhomologous). Of the deletion mechanisms diagrammed in Figure 3.7, the first (intrachromatid exchange) is limited to deletion, whereas the second (non–sister chromatid exchange) and third (sister chromatid exchange) give rise to duplication and deletion products in equal numbers. The fact that deletion occurs 10 times more frequently than duplication suggests that most deletions occur by the looping-out process of intrachromatid exchange, and not by unequal sister or non–sister chromatid exchange. Of course, it is possible that other factors influence the observed frequencies of duplication and deletion. For example, duplications may be generally less deleterious than the corresponding deletions, and duplications that cause only subtle changes in phenotype might remain undetected in the population.

There are several indications that some and perhaps many of the chromosomal rearrangements in the *DMD* gene occur in somatic and/or germ-line diploid cells prior to meiosis. This evidence, in the case of deletions, comes from the families in which two or more affected boys are found to have a deletion that is not present in the blood lymphocytes of the mother. These mothers must be germ-line mosaics whose deletion arose in a diploid cell during development. In terms of parental origin it appears that deletions arise in both males and females, with no strong bias toward either one.

Partial gene duplication is rarely described in other genetic disorders and its study in the *DMD* gene has provided insight into duplication mechanisms. In a study of the origin of five tandem duplications in our laboratory, four originated by chromatid exchange in the single X chromosome of the maternal grandfather and the fifth originated in a single X chromosome of the maternal grandmother. Thus, all five duplications arose by an intrachromosomal event such as unequal sister chromatid exchange. Since the grandfathers were all unaffected, the rearrangements are presumed to have taken place in the germ line, but whether they took place during a mitotic division cycle or during meiosis is not clear. The fact that duplications appear to be more frequently of male origin may reflect the higher number of mitotic divisions involved in generation of male

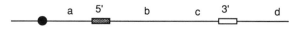

NORMAL CHROMOSOME

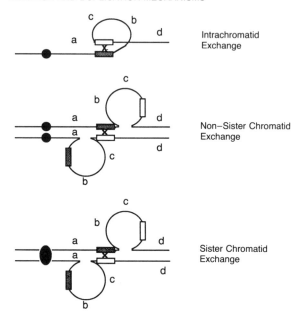

DELETION AND DUPLICATION MECHANISMS

Intrachromatid
Exchange

Non–Sister Chromatid
Exchange

Sister Chromatid
Exchange

DUPLICATION AND DELETION PRODUCTS

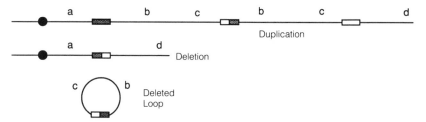

Duplication

Deletion

Deleted
Loop

Figure 3.7: Schematic of deletion and duplication mechanisms. The shaded and open boxes represent 5′ and 3′ elements involved in the recombination event. A representation of the normal chromosome is shown at the top of the figure. Below this are shown three mechanisms for the generation of deletions and/or duplications. Duplication and deletion end products are shown at the bottom of the figure. The first mechanism (intrachromatid exchange) results in the formation of the deletion chromosome and the deleted loop. The other two mechanisms are similar in nature, except that one is an exchange between chromatids of homologous chromosomes and the other is between chromatids of the same chromosome. Both these recombination events result in one chromosome carrying the duplication and one carrying the deletion. The recombination events portrayed in all three diagrams will be classified as homologous recombination if the 5′ and the 3′ elements have the same, or similar, sequence; they will be classified as nonhomologous recombination if these two elements have unrelated sequences.

germ cells or the possibility of an increased frequency of meiotic intrachromosomal exchange in the absence of a homologous X chromosome. Paternal origin of the mutation has also been observed for eight out of eight X-autosome translocations ascertained in affected females. This is unlikely to be attributable to increased opportunity for mitotic error in male gametogenesis, however, since male carriers of X-autosome translocation are often sterile due to spermatogenic arrest at meiosis. Thus, translocations arising in the germ line are unlikely to undergo successful meiosis, implying a meiotic origin for these translocations. The exclusively paternal origin for these translocations might possibly be due to a greater propensity to undergo rearrangement in the absence of the homologous chromosome, as suggested above to explain the sex bias for duplications.

For deletions in the *DMD* gene, little information is yet available on the molecular details at the site of the chromosomal rearrangements. To date none of the deletion breakpoints have been sequenced to see if the exchanges are between interspersed repeat units or between unrelated 5' and 3' elements (Figure 3.7). This type of study will be particularly important for the deletions that occur in the hot spot near the middle of the gene, for one might anticipate that they will be caused by a high frequency event peculiar to that region of the chromosome. Sequencing through breakpoints will determine if the exchanges are between homologous repeats or between nonhomologous regions of the X chromosome. In the case of the latter, one might assume that the high frequency of exchanges will be related to some discernible feature of the chromosome structure at that region of the chromosome.

Sequencing through duplication junctions is also important to determine if the exchanges are of the homologous or the nonhomologous type. A study of three duplication junctions in our laboratory has shown that one duplication arose by homologous recombination between Alu sequences while two others arose by a nonhomologous exchange mechanism (Hu, personal communication).

Three translocation breakpoints have also been sequenced in our laboratory. All three were shown to involve nonhomologous recombination, and in one the tetranucleotide sequence CGGC occurred several times near the exchange site, implicating it as a possible potentiator sequence or recognition sequence for enzymes involved in the translocation process.

MOLECULAR MEDICINE

Diagnostic Applications

Now that the *DMD* gene has been cloned and the gene product can be visualized, the diagnosis of DMD and BMD is most readily confirmed by visualizing the underlying defect directly, at the level of either the *DMD* gene or the protein it encodes. This is particularly useful for differential diagnosis of Becker muscular dystrophy from other muscle disorders such as limb girdle muscular dystrophy or spinal muscular atrophy, the distinction being especially difficult for sporadic cases with no family history.

DNA Analysis—Southern Blot and PCR Amplification   Partial deletions are readily visualized by Southern analysis with *DMD* cDNA probes. The *DMD* gene consists of at least 70 exons detected by a variety of partial clones of the 14 kb cDNA. These were described above and in Figure 3.1. Generally, Southern analysis is performed on *Hin*dIII digests, as the order of the exon-containing fragments is best defined for this enzyme. Confirmation using a second enzyme (*Pst* I, *Eco*RI, *Bgl* II, or *Taq* I), has proven to be useful to detect deletions of restriction fragments that comigrate with other fragments on *Hin*dIII digests. Quantitation of Southern blots by scanning laser densitometry permits the detection of female deletion carriers who have one copy of the deleted region, and also of males with partial gene duplication who have two copies of the duplicated region.

Generally, predictions as to the effect on the translational reading frame may be made on the basis of Southern analysis to identify deleted or duplicated exons, and reference to Table 3.1 to determine the most likely splicing patterns. Patients whose deletions or duplications are of the frame-shift type are much more likely to have a severe phenotype than those with genetic lesions that are expected to maintain the translational reading frame. Of course there are exceptions to this rule, as discussed above and in Table 3.2. Patients with a unique junction fragment detected by cDNA probes are excluded from this predictive analysis, since such fragments may be associated with deletion extending into the exon.

The finding of a novel junction fragment associated with deletion or duplication facilitates carrier identification since the bands generated by these fragments are readily observed without resorting to quantitative analysis. Although a junction fragment must be generated by every deletion or tandem duplication, many are not detected in a routine analysis since only a portion of the *DMD* gene is detected by cDNA probes and genomic probes from the cloned regions of the gene. Hybridization of Southern blots with whole cosmid DNA has been used to increase the number of detectable junction fragments over and above that detected by currently available probes.

Most mutations detectable by standard Southern analysis are also detectable by pulsed field gel electrophoresis (PFGE). This technique resolves fragments from 50 kb to several megabases in size, and will detect deletions, duplications, and complex rearrangements if these result in changes in fragment size of greater than about 30–40 kb. PFGE allows visualization of a large portion of the gene with a single probe, and it also detects many duplications and deletions as novel-size fragments, again obviating the need for quantitative analysis to detect the mutant gene in carriers. Furthermore, PFGE has demonstrated its potential to detect more complex rearrangements, such as insertions that were not detected by standard Southern analysis.

Deletions may also be detected by the polymerase chain reaction (PCR). Primer pairs have been designed to amplify specific exons and the presence or absence of each exon scored by ethidium bromide staining without the need for radioisotope-based techniques. Primer pairs representing the most commonly deleted exons have been combined in one "multiplex" amplification capable of scoring approximately 80% of DMD and BMD deletions. A complementary primer set has been described, which when combined with that of Chamberlain permits detection of over 98% of those deletions detectable by Southern analysis with cDNA probes. Thus, nearly all deletions may be identified in two PCR reactions readily performed in a single day. Prognosis on the basis of predicted effect on translational reading frame, however, does require subsequent Southern analysis to delineate the extent of the deletion. Conceivably, suitable primer sets might be derived to "speed up" this secondary analysis as well.

An alternative approach is amplification of reverse-transcribed mRNA sequences (RT-PCR) from muscle or from "illegitimate" transcripts of blood lymphocytes. The utility of this approach has been demonstrated by Chelly et al. and more recently by Roberts et al. in several patients with DMD gene rearrangements.

Dystrophin Analysis—Western Blot and Immunocytochemistry Dystrophin analysis requires a muscle biopsy, usually a portion of the biopsy taken for routine pathological diagnosis. Dystrophin analysis has the distinct advantage of allowing direct detection of defects in the DMD gene product, irrespective of knowledge about the underlying mutation; its major disadvantages are that it is not practical for prenatal diagnosis and it is not yet reliable for carrier identification. Also, severity of phenotype is likely to be related to both qualitative and quantitative changes in dystrophin, yet neither the levels of dystrophin necessary to prevent muscle weakness nor the essential functional domains of dystrophin have been adequately defined.

Hoffman et al. analyzed Western blots of muscle from 56 Duchenne and Becker patients and reported that nearly all Duchenne patients exhibited no detectable dystrophin (< 3%), while more than half of intermediate or Becker patients exhibited detectable dystrophin of abnormal size. In a subsequent study of 97 patients with a possible diagnosis of BMD, approximately half had dystrophin of altered molecular weight while 10% had normal-sized dystrophin estimated at 5–30% of the normal concentration. The remaining 40%, one-third of whom were female, had normal-size and near-normal levels of dystrophin, consistent with a functionally important but undetectable lesion in the *DMD* gene. Alternatively, it is likely that some of these were misdiagnosed, especially the females and the males without a clear-cut X-linked family history. Based on these studies it has been suggested that patients might be classified into one of three groups: (1) DMD, wheelchair-bound by about age 11—dystrophin level < 3% of normal; (2) severe BMD, wheelchair-bound at 13–20 yrs—dystrophin level 3–10% of normal; (3) moderate or mild BMD, wheelchair-bound after 20 yrs— dystrophin level ≥ 20% of normal.

Recent results from two laboratories suggest that the above classification is inaccurate and quite misleading. Nicholson et al. have examined 140 DMD/BMD patients by Western blot analysis and by immunofluorescence on muscle sections and have concluded that overall the estimated abundance of dystrophin correlated well with the clinical assessment of disease severity. The levels of detectable dystrophin were considerably higher than those estimated by Hoffman, and more importantly the majority of severe DMD patients had detectable dystrophin of altered molecular weight with abundance of 8–40% of normal. In this study, a highly specific monoclonal antibody was used, and all Western blot results were quantitated by scanning densitometry. All dystrophin amounts were normalized to myosin levels to correct for differences in the amount of muscle tissue in different biopsies. In the Hoffman studies the results were visual estimates instead of instrument quantitation. In our laboratory we have chosen to quantitate both myosin and dystrophin by scanning laser densitometry of stained gels and Western blots, respectively. Furthermore, we have used antibodies against both the carboxy-terminal and amino-terminal regions of the protein. In our initial study, five of seven severe DMD cases were found to have an altered-size dystrophin with staining intensity of 10–17% of normal levels, detected with the amino-terminal but not the carboxy-terminal antibody. This is in contrast to the results from Hoffman, which suggested that truncated gene products, lacking the normal carboxy terminus, are rapidly degraded. The Becker patients in our study had detectable dystrophin of altered size, which reacted with both amino-terminal and carboxy-terminal antibodies, consistent with a deletion that maintains the translational reading frame. The carboxy-terminal antibody may therefore be of value in providing a differential diagnosis between BMD and DMD. Further study is required to determine if the truncated protein products detected in DMD patients are of functional significance and also to examine the correlation between dystrophin levels and phenotypic severity.

Immunocytochemical staining of the surface membrane with antibodies against dystrophin is a more sensitive technique than that of Western blotting and it examines the localization of dystrophin to the sarcolemma, assumed to be crucial for its biological function. Dystrophin staining is absent or markedly deficient in DMD patients as well as in *mdx* mice and has been described as "sporadic" or "diffuse" in BMD patients. While immunocytochemical analysis may provide definitive diagnosis in dystrophin-negative biopsies, a "patchy" staining may be difficult to interpret, especially since this type of staining has been seen in other muscle disorders (e.g., polymyositis). DNA analysis coupled with the dystrophin analysis would be of benefit in the interpretation of such results.

Symptomatic carriers have also been shown to have a "mosaic" or "patchy" pattern of immunological staining, with groups of positive and negative fibers and relatively few partially staining fibers. Obligate carriers who do not manifest the disease have been reported to have a much lower proportion of negatively stained fibers such that reliable detection of obligate carriers by this technique may not be feasible. Arahata et al. have suggested that elevated CK levels might result from segmental dystrophin deficiency in individual muscle fibers, suggesting that carriers with elevated CK levels will be detectable by immunofluorescence.

Prevention—Carrier Identification and Prenatal Diagnosis

Prior to the discovery of genetically linked probes in 1983, carrier status was determined by testing for elevated levels of creatine kinase in the serum, an inadequate test since only 70–75% of obligate carriers have elevated CK levels. For obligate carriers or for females identified to be at risk by means of the CK test, prenatal diagnosis was limited to fetal sexing by karyotype analysis, with no means of distinguishing an affected from an unaffected male.

Genetic Markers for Carrier Identification and Prenatal Diagnosis   The application of molecular markers (RFLPs) to carrier detection, with the ability to distinguish the region of the X chromosome carrying the DMD mutation from the homologous region on the "normal" chromosome, was first reported in 1983 using the first markers flanking the *DMD* gene. The accuracy with which carrier status can be determined is greatly enhanced using a pair of flanking markers, the error determined by the frequency of double recombination between the markers. Using the full set of genetic markers available in 1986, carrier status could be predicted with over 98%

accuracy for the 75% of cases informative with markers flanking the gene. When an intragenic marker is also informative, recombination between the flanking markers is readily detected, and in cases where no cross-over is apparent, diagnostic error is reduced to a few tenths of a percent. The 25% of cases for whom accurate carrier detection is not possible include those with detectable recombination between the closest flanking markers, cases where informative markers are only found to one side of the DMD gene, and cases in which phase (assignment of the normal and mutant alleles to each of the two X chromosomes) cannot be determined.

Application of molecular techniques to prenatal diagnosis requires fetal DNA derived from either chorionic villi or cultured cells from amniotic fluid and generally permits diagnosis of affected males with an accuracy similar to that for carrier identification. In the 60–70% of families for which the DMD mutation is defined as a deletion or duplication, presence or absence of the mutation in the fetal DNA is scorable directly using Southern analysis, polymerase chain reaction, or, more rarely, PFGE. In families for which the DMD mutation is undefined, prenatal diagnosis is dependent on linkage analysis using informative intragenic and flanking markers. In this case, accuracy of diagnosis is dependent on the same criteria and subject to the same uncertainties as those involved in carrier detection. It has been estimated that, based on linking data without recourse to cDNA analysis, 90% of carrier women have the potential for prenatal diagnosis with 95% or greater accuracy.

For most genetic diseases, one intragenic probe used alone would provide accurate diagnostic information, since the frequency of recombination between the site of the mutation and the site detected by the probe is expected to be essentially zero. In DMD, however, the mutation site for any given family is often unknown, and since the gene is so large, recombination may occur between an informative intragenic marker and the site of the mutation. Initially, family studies with the PERT87 and XJ probes gave an empirical estimate of 5% intragenic recombination between these markers and the mutations segregating in the families. Diagnostic evaluation based on these markers was therefore assumed to be 95% accurate. Not surprisingly, two false-negative prenatal diagnoses resulting from such intragenic recombination have been reported.

The empirical estimate of intragenic recombination frequency of 5% is higher than would be expected for the length of the gene. This estimate was suspected to be artificially high due to the occurrence of germ-line mosaicism, as described below. Male siblings who were discordant for DMD but concordant for intragenic markers, had been scored as being due to recombination. However, such discordancy might often be due to a mother who is a germ-line mosaic, carrying both normal and mutant alleles on the same genetic background but in different germ-cell lines, resulting in an artificially high estimation of recombination frequencies. A recent study by Abbs et al. using polymorphic markers at the two extremities of the gene gave an estimated total intragenic recombination frequency of nearly 12% with confidence limits of 4–23%, which suggests that the observed recombination frequency levels are *not* artificially elevated.

For familial cases of DMD, information provided by linkage analysis tends to shift prior carrier risk estimates based on CK toward the extremes, such that individuals at lower risk often are given a much lower revised estimate after the analysis, while individuals at greater risk often have a still higher revised estimate. Relatively few individuals end up with intermediate risks of 20–80%. For DMD occurring in families with no family history, however, linkage analysis cannot distinguish a mother who carries the mutation herself from one whose son is a new mutant.

Germ-line Mosaicism An affected boy who presents as the first affected individual in his family, whose mutation is defined, and whose mother does not carry the mutation in her lymphocytes, has generally been considered to represent a new mutation at the *DMD* locus. On this basis, his mother has not been considered to be at risk for a second affected son. Since 1986, however, a number of reports have shown recurrence of the same "new mutation" in multiple siblings within a nuclear family. The second occurrence of the same "new mutation" is most readily explained if the mother is mosaic for the mutation in her germ line. Similarly, male germ-line mosaicism may result in the appearance of two carrier daughters with the same "new mutation" in a family where neither parent shows the mutation in lymphocyte DNA.

Germ-line mosaicism greatly complicates determination of carrier status, since the mother of *any* sporadic case might be a germ-line mosaic. Such an individual would not have an elevated CK, nor would her son's mutation, if known, be present in her own lymphocyte DNA, yet the risk of having a second affected son or a carrier daughter is dependent on the proportion of her germ cells carrying the mutation. Based on their own patient group, Bakker et al. determined the empiric recurrence risk to be 14% for a male fetus sharing the same X chromosome markers as an affected sibling in whom a "new mutation" had previously been identified.

Pathophysiology of the Duchenne and Becker Muscular Dystrophies

Despite decades of intensive research into the cause of Duchenne muscular dystrophy, the nature of the basic defect remained elusive until the cloning of the gene allowed the identification of the dystrophin molecule as the defective protein. Even with this new knowledge, however, the biological role of dystrophin remains speculative and our understanding of the disease remains incomplete. Certainly there are many clues from the earlier literature that should be re-examined in the light of the new knowledge.

In the history of muscular dystrophy research, a number of models for the disease pathophysiology have been considered. These have included (1) vascular disruption or (2) defective innervation leading to muscle degeneration, (3) increased myoplasmic Ca^{2+}, triggering activation of proteases to cause muscle breakdown, and (4) a defect in protein synthesis or degradation leading to a generalized wasting of the muscle tissue. All of these ideas have suffered from inconsistent and sometimes contradictory evidence, in many cases a result of the difficulty in distinguishing the primary defect from the many secondary manifestations of the disease, in a tissue that is undergoing degeneration and replacement by fat and connective tissue.

The possibility of a generalized membrane defect is one outcome of the earlier research that has maintained continued support. This defect is expressed most clearly in muscle tissue in vivo but also, under certain conditions, is apparent in other cell types both in vivo and in vitro. In brief, the evidence includes (1) the finding of gaps or lesions in the plasma membrane in electromyographic studies of prenecrotic muscle tissue from affected boys, (2) the finding of greatly increased levels of certain muscle enzymes in the serum of young presymptomatic boys, suggesting leakage of macromolecules through the muscle membrane, (3) an increased level of Ca^{2+} in muscle fibers possibly due to increased uptake through a "leaky" membrane, (4) alterations in lectin binding to glycoproteins on the muscle cell surface, and (5) apparently altered intercellular adhesiveness of skin fibroblasts from DMD patients. These studies served to illustrate the concept of a generalized membrane defect but fell short of determining the biochemical process or the protein component that is directly affected by the mutation in the *DMD* gene.

An alteration in the muscle membrane is quite consistent with the recent studies that have identified dystrophin as the product of the *DMD* gene and determined it to be a high molecular weight cytoskeletal protein localized at the sarcolemmal membrane. An attractive working model is one in which dystrophin plays a direct and fundamental role in muscle membrane stability. According to the model, the basic defect in DMD is the lack of functional dystrophin, which results in a weakened membrane that is susceptible to contraction-induced tearing. In boys with BMD, perhaps, the presence of a reduced amount of normal dystrophin or of partially functional dystrophin gives partial stability to the membrane and results in a milder phenotype. In both DMD and BMD the localized membrane lesions would be expected to give rise to the segmental necrosis that is observed in the early stages of the disease. This phase is followed by degeneration and regeneration of these segments, regeneration proceeding through the proliferation and differentiation of satellite cells (myoblasts) that move in to repair the damage. In the later stages of the disease, the regenerative capacity would be expected to decline as the finite proliferative potential of the satellite cells is used up and the satellite cells themselves become depleted. This view of the disease progression is consistent with the reduced growth potential of myoblasts derived from the muscle tissue of affected boys.

In further support of the model is the previously mentioned evidence that dystrophin is part of a complex with five other proteins. One of these, the 156-kD glycoprotein, dystroglycan, binds laminin on the outside of the membrane. Furthermore, the presence of an actin-binding domain near the N-terminal end of the dystrophin molecule (if verified) would suggest that the dystrophin-glycoprotein complex might serve to connect the internal cytoskeleton of actin filaments to the basal lamina on the outside of the membrane. It is not difficult to imagine how such a complex might protect the membrane from damage during muscle contraction and relaxation. Verification of this model will await confirmation of the actin-binding capacity of dystrophin and a more detailed description of the proteins that interact with dystrophin at the cell surface.

In continuing the study of the basic defect, it is important not to become wed to any particular model. In this regard we recall the suggestion of Campbell and Kahl that the biological role of dystrophin might be to maintain a nonuniform distribution of an associated membrane glycoprotein on the surface of the myofiber, where it might function as an ion channel or a membrane receptor. Potentially important in this regard are two groups' recent reports of alterations in the nongated Ca^{2+} channels in DMD human and *mdx* mouse muscle. These reports reopen the idea that increased cytoplasmic Ca^{2+} may stimulate protease activity, resulting in proteolysis of muscle tissue. This would, at the moment, appear to be an alternative to the mechanical disruption model described above.

Whichever model turns out to be correct, it should not take very long for the correct physiological role of dystrophin to emerge. Once this is known, the major task will be to develop therapeutic strategies based on these new insights.

Therapeutic Options and Directions

Even before the detailed analysis of dystrophin is complete, there are a number of approaches that are being explored in an attempt to develop an effective therapy for Duchenne and related dystrophies. These include (1) the continued development of animal models of the disease, (2) the testing of prednisone therapy in patients, (3) the initiation of myoblast transplantation in animals and in patients, and (4) the development of newborn screening technology for the early identification of affected boys who might benefit from the therapies that are introduced.

Animal Models   Models of muscular dystrophy have been studied extensively in the mouse, hamster, chicken, and more recently in the dog. The murine X-linked form of muscular dystrophy is the *mdx* mouse. The genetic defect has recently been shown to be a point mutation in the mouse "*DMD* gene" that changes an amino acid codon to a stop codon, resulting in premature termination of the dystrophin molecule. The shortened protein fails to localize at the membrane and may be rapidly degraded, as evidenced by the complete lack of dystrophin staining in immunostained sections of *mdx* muscle.

While the *mdx* mouse is a good genetic model for DMD, the phenotype is considerably different from the human disease. In particular, *mdx* muscle has substantial powers of regeneration, resulting in the restoration of a relatively normal muscle morphology and function by about four weeks of age. The lack of a severe phenotype in the absence of dystrophin suggests that dystrophin is not required for membrane stability in mouse muscle. One possibility is that stress on the membrane is related to the fiber size and dystrophin may be required for the integrity of large diameter fibers in human muscle but not for the smaller diameter fibers in mice. This would not, however, explain the substantial *mdx* muscle degeneration in the neonatal period, nor is it consistent with the extensive conservation of the large gene throughout evolution.

Another possibility is the existence of a dystrophin-like protein that is encoded by a gene at a different locus and that is capable of substituting for dystrophin in the membrane of mouse muscle. Induction of this hypothetical protein during the first round of degeneration in the *mdx* mouse would be necessary to explain the subsequent stabilization of the phenotype. A dystrophin-like protein identified in Davies' laboratory is related to dystrophin by sequence homology at the C-terminal end and is encoded by a gene on human chromosome 6. Although there is no evidence to suggest a role for this protein in the modification of the *mdx* phenotype, the existence of a dystrophin-like family of proteins does leave open this possibility.

For many types of experimentation it would be useful to have a more faithful model of the human disease. Fortunately, such a model exists in the canine X-linked muscular dystrophy (CXMD). This disease, discovered in the golden retriever, faithfully mimics the phenotype of Duchenne muscular dystrophy, and the affected animals fail to show dystrophin staining by

Western blot analysis or in cryostat sections. The affected animals are also deficient in *DMD* mRNA and the mutation in the *DMD* gene has recently been characterized. The dog model of DMD may be of great value in the evaluation of new therapeutic strategies, especially in the development of myoblast transplantation and gene therapy as described below.

Prednisone Therapy   For several years prednisone has been reported to have a potential short-term benefit for boys with DMD. Recently, a six month randomized double-blind trial comparing 0.75 and 1.5 mg/kg of prednisone (daily) with a placebo group has revealed significant improvement at both doses in several parameters relating to muscle function and total muscle mass. This improvement was apparent after one month, continued until three months and remained constant thereafter until three years. Although the results were encouraging, dramatic improvement was lacking; joint contractures remained and those requiring leg braces or wheelchairs at the start of the study still relied on these supports at the end of the study.

The mechanism by which prednisone may act to produce these results is unknown. Steroid hormones can act as gene regulatory molecules, suggesting that a steroid-receptor complex could stimulate transcription of the *DMD* gene to give increased concentrations of a partially functional dystrophin molecule. This idea is readily testable with the current tools. Another possibility is that the beneficial effects of prednisone are mediated through an anti-inflammatory effect that slows the process of muscle fiber necrosis. Whatever the mechanism, Mendell et al. caution that prednisone is not advocated as a treatment for DMD. The long-term risks of prednisone act as a deterrent to its widespread use and suggest that the encouraging results should be viewed as an initial step toward identification of more satisfactory therapeutic agents.

Myoblast Transplantation and Gene Therapy   With the finding that the *DMD* gene product is a very large cytoskeletal protein came the realization that treatment of the disease through direct replacement of the protein will be difficult, if not impossible. Therefore, introduction of a new dystrophin gene into the muscle appears to be the best alternative at the present time. There are potentially two routes to achieve this end. One is myoblast transplantation, in which myoblasts from a normal donor are injected into the dystrophin-deficient muscle, where they are expected to fuse with and become part of the host tissue. The second is gene therapy, which, in its simplest form, would involve removal of patient myoblasts, introduction of a new dystrophin gene into the cells, and reimplantation of the transfected cells into the muscle tissue.

Myoblast transplantation to correct the defect in DMD was the subject of an international workshop sponsored by the Muscular Dystrophy Association in June of 1989. At that meeting it became clear that (1) the technical difficulties involved in the transplantation approach are considerable, perhaps even formidable, (2) the animal studies to date are promising but fall far short of answering all the questions that need to be answered, and (3) some questions can only be answered in humans, and for this reason a few groups were contemplating limited transplantation studies in patients.

Skeletal muscle is almost the ideal issue for cell transplantation. During development, myoblasts fuse to form a multinucleated fiber. Even in mature muscle, a pool of myoblasts termed "satellite cells" do not fuse but remain as single myogenic cells lying in the sarcolemma between the plasma membrane and the basal lamina. These cells are capable of fusing to one another and to the existing muscle to regenerate muscle fibers following injury or disease. In DMD the satellite cells become depleted after several rounds of degeneration and regeneration, and the introduction of donor cells capable of fusing to the existing muscle and carrying an intact *DMD* gene is an attractive possibility.

The impetus for myoblast transplantation has come from pioneering studies in the mouse. Building on a background of transplantation research by Partridge and his colleagues, two groups have successfully transplanted normal myoblasts into a single muscle of *mdx* mice and demonstrated that the donor cells fuse with the existing muscle and produce dystrophin. In Partridge's study the dystrophin was localized to the membrane, although it had a patchy distribution and the amount of dystrophin on Western blot analysis was as high as 30–40% of normal levels. Similar results were obtained by Karpati et al. following the transplant of normal human myoblasts into mouse *mdx* muscle.

Although the *Dy* gene is not the mouse equivalent of the human *DMD* gene, Law has carried out transplantation into multiple muscle groups in the *dy/dy* mouse, with an autosomal recessive form of dystrophy, and has claimed significant functional recovery following transplantation. This result clearly needs to be confirmed and extended.

To a large extent, transplantation studies are in their infancy, and many unanswered questions remain. Many of these will be difficult to answer in the mouse, and will be better addressed in the dog or in affected boys. Pilot studies injecting a single muscle to test the extent of immune rejection are already being done in affected children. On the other hand, extensive experimentation involving the injection of multiple muscle groups, whose end point is the demonstration of improved muscle function, will best be done in the dog. In dogs, for example, it should be possible to determine the best routes of injection, the number of injection tracts, the number of cells to be injected per tract, and the timing of the injections for the major muscle groups. It should also be possible to determine the extent to which the injected cells migrate through the muscle, the proportion of donor cells that fuse immediately and the proportion that remain as satellite cells for future use, and the level of dystrophin production necessary to achieve a clinical improvement.

Most of these questions would require at least a partial answer in animals before young patients are subjected to the process of multiple injections. In the meantime, however, at least five groups have decided to proceed with a limited human trial to determine if the injection of myoblasts into one muscle is tolerated without severe reaction, to determine if the donor nuclei will be stably incorporated into the host muscle without rejection, and to determine if dystrophin is produced in the transplanted muscle. Two preliminary reports of dystrophin production in a single patient is all that has been published to date.

If the dog experiments prove promising and if human studies indicate that injected cells are tolerated without untoward effects, then transplantation may become a viable option for young boys with muscular dystrophy. The procedure will be far from trivial, however. It is not clear if one set of injections will be sufficient or if repeated injections throughout life will be required. Finally, there will still be the problem of the heart, which likely will not be amenable to transplantation therapy since cardiac muscle is not a syncytium. It is possible that the successful correction of the basic defect in the skeletal muscle will result in a significant improvement in the quality of life, but will put increased strain on the heart and result in the accelerated appearance of cardiomyopathy.

The alternative to donor myoblast transplantation is gene therapy. This approach is attractive for a number of reasons. First, it has the potential of using the patient's own myoblasts as a vehicle to introduce a new gene into the muscle, circumventing the immune rejection problem. Second, it offers the possibility of inserting a gene engineered to give enhanced dystrophin production, thereby allowing the use of fewer transplanted cells. Third, it may be possible to introduce a functional *DMD* gene directly into the muscle tissue, instead of using the patient's myoblasts as a vehicle, and if this proves effective it might then be possible to provide a new gene to the heart muscle.

The large size of the gene would prohibit gene therapy using an intact gene. Recently, a mouse "minigene" has been constructed that carries all the coding regions with the introns removed. This minigene is essentially a full-length cDNA driven by a non–muscle-specific promoter and it expresses dystrophin in nonmuscle cells in culture. This construct is approximately 15 kb, a size still rather large for gene transfer. Our own laboratory has taken a slightly different approach and constructed a smaller minigene, eliminating exons 16–48 from the construct and thereby removing a large segment of the spectrin-like repeat portion of the encoded dystrophin molecule. The fact that a very mildly affected male has been reported with a similar deletion suggests that this engineered human dystrophin molecule may be functional. The addition of the natural muscle-specific promoter of the *DMD* gene has allowed the minigene to be expressed and localized at the membrane in cultured myocytes from a dystrophin-negative patient.

Before gene therapy can be introduced there is still a need for considerable basic research. It will be essential to determine the most important functional domains of the dystrophin molecule, so that the corresponding regions of the gene can be retained in the minigene construct. Also it will be necessary to understand in detail the regulatory regions of the *DMD* gene so that a minigene might be introduced with a promoter engineered to give maximum transcriptional activity.

Newborn Screening   If any form of therapy proves to be of value it will likely be most effective if it is applied early in life. Since many affected boys are not diagnosed until the disease process is quite advanced, newborn screening may be the best way to guarantee that diagnosis will be early enough to maximize therapeutic value.

The technology for newborn screening has been available since the mid-1970s, utilizing a sensitive bioluminescence assay for the amount of creatine kinase in a dried blood spot. The assay has been applied to newborn screening in Germany, France, and Canada. The Canadian study is the only one that is compulsory and has screened virtually all newborns in the Province of Manitoba over a three-year period. Eight affected males were identified in approximately 27,000 screened, and five of these were confirmed with DNA probes for the *DMD* gene. The German study has run as a voluntary program for over 10 years, identifying 85 affected boys in 284,500 screened. In all the studies, the rate of false positives was very low, especially after the initial period during which cut-off values were established. The rate of false negatives is more difficult to determine without a substantial waiting period and a very thorough follow-up of the tested children. However, to date no cases of DMD have come to light that were not detected in the Manitoba screen, and the detection frequency of approximately 1 in 3,400 suggests that the false negative frequency is very low.

Screening, until now, has been done primarily for research purposes or as pilot projects designed to test the feasibility of mass screening. At the moment it is generally agreed that there is little justification to embark on government-legislated screening programs. All that could be offered to the families in whom an affected child is detected is early genetic counseling to prevent the birth of a second affected child. While this is a worthwhile aim, it has generally been accepted by the genetic and medical community that mandatory screening is warranted only when there is the possibility for successful intervention to alter the course of the disease in the affected child. The successful application of myoblast transplantation or any other form of therapy would, of course, satisfy this criteria.

CONCLUDING REMARKS

Our understanding of Duchenne and Becker muscular dystrophy has progressed in remarkable leaps in recent years. The identification of the *DMD* gene and its product, dystrophin, has given us substantial new insights into the basic defect in these diseases. Although the detailed understanding of the role of dystrophin is not yet in hand, the knowledge gained from the molecular biology approach clearly points the way to future experiments. The discovery of the responsible gene and protein has been frequently referred to as the "end of the beginning." Let us hope that dystrophin research, coupled with studies to evaluate therapeutic approaches, will mark the beginning of the end.

ACKNOWLEDGMENTS

The authors are grateful to Mr. Dennis Bulman for providing the Western blots and immunocytochemistry pictures in Figures 3.3B, C and D, to Dr. Michael Cullen for providing the electron micrograph used in Figure 3.3E, to Dr. Henry Klamut for providing Figure 3.4, and to Drs. Don Love and Kay Davies for information on their cDNA clones for Figure 3.1. We are also grateful to all members of our laboratory, especially Sharon Bodrug, Dennis Bulman, Xiuyuan Hu, and Henry Klamut for unpublished information.

REFERENCES

Abbs S, Roberts RG, Mathew CG, Bently DR, Bobrow M. Accurate assessment of intragenic recombination frequency within the Duchenne muscular dystrophy gene. Genomics 1990;7:602–606.

Arahata K, Hoffman EP, Kunkel LM, et al. Dystrophin diagnosis: comparison of dystrophin abnormalities by immunofluorescence and immunoblot analysis. Proc Natl Acad Sci USA 1989;86:7154–7158.

Bakker E, Veenema H, den Dunnen JT. Germinal mosaicism increases the recurrence risk for 'new' Duchenne muscular dystrophy mutations. J Med Genet 1989;26:87–93.

Bar S, Barnea E, Levy Z, Neuman S, Yaffe D, Nudel U. A novel product of the Duchenne muscular dystrophy gene which greatly differs from the known isoforms in its structure and tissue distribution. Biochem J 1990;272:557–560.

Baumbach LL, Chamberlain JS, Ward PA, Farwell NJ, Caskey CT. Molecular and clinical correlations of deletions leading to Duchenne and Becker muscular dystrophy. Neurology 1989;39:465–474.

Blau HM, Webster C, Pavlath FK. Defective myoblasts identified in Duchenne muscular dystrophy. Proc Natl Acad Sci 1983;80:4856–4860.

Boyce FM, Beggs AH, Feener C, Kunkel LM. Dystrophin is transcribed in brain from a distant upstream promoter. Proc Natl Acad Sci USA 1991;88:1276–1280.

Brooke MH, Fenichel GM, Griggs RC, et al. Duchenne muscular dystrophy: patterns of clinical progression and effects of supportive therapy. Neurology 1989;39:475–481.

Bulman DE, Murphy EG, Zubrzycka-Gaarn EE, Worton, RG, Ray P. Differentiation of Duchenne and Becker muscular dystrophy phenotypes with amino- and carboxy-terminal antisera specific for dystrophin. Am J Hum Genet 1991;48:295–304.

Burghes AHM, Logan C, Hu X, Belfall B, Worton RG, Ray PN. A cDNA clone from the Duchenne/Becker muscular dystrophy gene. Nature 1987;328:434–437.

Campbell KP, Kahl SD. Association of dystrophin and an integral membrane glycoprotein. Nature 1989;338:259–262.

Chamberlain JS, Gibbs R, Ranier JE, Nguyen PN, Caskey CT. Deletion screening of the Duchenne muscular dystrophy locus via multiplex DNA amplification. Nucl Acids Res 1988;16:11141–11156.

Chelly J, Gilgenkrantz H, Hugnot JP, et al. Illegitimate transcription. Application to the analysis of truncated transcripts of the dystrophin gene in non-muscle cultured cells from Duchenne and Becker patients. J Clin Invest 1991;88:1161–1166.

Cooper BJ, Winand NJ, Stedman H, et al. The homologue of the Duchenne locus is defective in X-linked muscular dystrophy of dogs. Nature 1988;334:154–156.

Davies KE, Pearson PL, Harper PS, et al. Linkage analysis of two cloned DNA sequences flanking the Duchenne muscular dystrophy locus on the short arm of the human X chromosome. Nucl Acids Res 1983;11:2303–2313.

den Dunnen JT, Bakker E, Van Ommen GJB, Pearson PL. The DMD gene analysed by field inversion gel electrophoresis. Br Med Bull 1989;45:644–658.

Dubreuil RR. Structure and evolution of the actin crosslinking proteins. Bioessays 1991;13:219–226.

Emery AEH. Duchenne muscular dystrophy. Oxford monographs on medical genetics. No. 15. Oxford: Oxford University Press, 1987.

Ervasti JM, Campbell KP. Membrane organization of the dystrophin-glycoprotein complex. Cell 1991;66:1121–1131.

Feener CA, Koenig M, Kunkel LM. Alternative splicing of human dystrophin mRNA generates isoforms at the carboxy terminus. Nature 1989;338:509–511.

Fenichel GM, Florence JM, Pestronk A, et al. Long-term benefit from prednisone therapy in Duchenne muscular dystrophy. Neurology 1991;41: 1874–1877.

Fong P, Turner PR, Denetclaw WF, Steinhardt RA. Increased activity of calcium leak channels in myotubes of Duchenne human and *mdx* mouse origin. Science 1990;250:673–676.

Gillard EF, Chamberlain JS, Murphy EG, et al. Molecular and phenotypic analysis of patients with deletions within the deletion-rich region of the DMD gene. Am J Hum Genet 1989;45:507–520.

Hodgson SV, Bobrow M. Carrier detection and prenatal diagnosis in Duchenne and Becker muscular dystrophy. Br Med Bull 1989;45:719–744.

Hoffman EP, Brown RH, Kunkel LM. Dystrophin: the protein product of the Duchenne muscular dystrophy locus. Cell 1987;51:919–928.

Hoffman EP, Fischbeck KH, Brown RH, et al. Characterization of dystrophin in muscle-biopsy specimens from patients with Duchenne's or Becker's muscular dystrophy. N Engl J Med 1988;318:1363–1368.

Hoffman EP, Kunkel LM. Dystrophin abnormalities in Duchenne/Becker muscular dystrophy. Neuron 1989;2:1019–1029.

Hu X, Murphy EG, Ray PN, Thompson MW, Worton RG. Duplicational mutation at the Duchenne muscular dystrophy locus: its frequency, distribution, origin, and phenotype/genotype correlation. Am J Hum Genet 1990;46:682–695.

Ibraghimov-Beskrovnaya O, Ervasti JM, Leveille C, Slaughter CA, Sernett SW, Campbell KP. Primary structure of dystrophin-associated glycoproteins linking dystrophin to the extracellular matrix. Nature 1992;355:696–702.

Karpati G, Pouliot Y, Zubrzycka-Gaarn E, et al. Dystrophin is expressed in mdx skeletal muscle fibers after normal myoblast implantation. Am J Pathol 1989;134:27–32.

Klamut HJ, Gangopadhyay SB, Worton RG, Ray PN. Molecular and functional analysis of the muscle-specific promoter region of the Duchenne muscular dystrophy gene. Mol Cell Biol 1990;10:193–205.

Koenig M, Beggs AH, Moyer M, et al. The molecular basis of Duchenne versus Becker muscular dystrophy: correlation of severity with type of deletion. Am J Hum Genet 1989;45:498–506.

Koenig M, Monaco AP, Kunkel LM. The complete sequence of dystrophin predicts a rod-shaped cytoskeletal protein. Cell 1988;53:219–228.

Law PK, Bertorini TE, Goodwin TG, et al. Dystrophin production induced by myoblast transfer therapy in Duchenne muscular dystrophy. Lancet 1990;336:114–115.

Law PK, Goodwin TG, Li H-J. Histoincompatible myoblast injection improves muscle structure and function of dystrophic mice. Transplantation Proc 1988;20(suppl 3):1114–1119.

Lee CC, Pearlman JA, Chamberlain JS, Caskey CT. Expression of a recombinant dystrophin and its localization to the cell membrane. Nature 1991;349: 334–336.

Love DR, Morris GE, Ellis JM, et al. Tissue distribution of the dystrophin-related gene product and expression in the mdx and dy mouse. Proc Natl Acad Sci USA 1991;88:3243–3247.

Malhotra SB, Hart KA, Klamut JH, et al. Frameshift deletions in patients with Duchenne and Becker muscular dystrophy. Science 1988;242:755.

Mendell JR, Moxley RT, Griggs RC, et al. Randomized, double-blind six-month trial of prednisone in Duchenne's muscular dystrophy. N Engl J Med 1989;320:1592–1597.

Monaco A, Bertelson CJ, Middlesworth W, et al. Detection of deletions spanning the Duchenne muscular dystrophy locus using a tightly linked DNA segment. Nature 1985;316:842–845.

Nicholson LVB, Johnson MA, Gardner-Medwin D, Bhattacharya S, Harris JB. Heterogeneity of dystrophin expression in patients with Duchenne and Becker muscular dystrophy. Acta Neuropathol 1990;80:239–250.

Nudel U, Zuk D, Einat P, et al. Duchenne muscular dystrophy gene is not identical in muscle and brain. Nature 1989;337:76–78.

Partridge TA, Morgan JE, Coulton GR, Hoffman EP, Kunkel LM. Conversion of mdx myofibres from dystrophin-negative to -positive by injection of normal myoblasts. Nature 1989;337:176–179.

Ray PN, Belfall B, Duff C, et al. Cloning of the breakpoint of an X;21 translocation associated with Duchenne muscular dystrophy. Nature 1985;318:672–675.

Roberts RG, Barby TF, Manners E, Bobrow M, Bentley DR. Direct detection of dystrophin gene rearrangements by analysis of dystrophin mRNA in peripheral blood lymphocytes. Am J Hum Genet 1991;49:298–310.

Sicinski P, Geng Y, Ryder-Cook AS, Barnard EA, Darlison MG, Barnard PJ. The molecular basis of muscular dystrophy in the *mdx* mouse: a point mutation. Science 1989;244:1578–1580.

Witkowski JA. Dystrophin-related muscular dystrophies. J Child Neurol 1989;4:251–271.

Worton RG, Thompson MWT. Genetics of Duchenne muscular dystrophy. Ann Rev Genet 1988;22:601–629.

Zubrzycka-Gaarn EE, Bulman DE, Karpati G, et al. The Duchenne muscular dystrophy gene product is localized in the sarcolemma of human skeletal muscle. Nature 1988;333:466–469.

CHAPTER 4

Huntington's Disease

James Gusella
Anne B. Young

HISTORY AND CLINICAL MANIFESTATIONS

Dr. George Huntington first provided a succinct description of this disease that bears his name in 1872. He had observed as a child, and later as a physician on Long Island, several families that had an inherited disorder of abnormal involuntary movements and disturbed behavior. In his classic report he described most of the salient features of the disease as we know it today.

Huntington's disease (HD) is a neurodegenerative disorder that is inherited in an autosomal dominant fashion. The average age of onset is between 35 and 40 years of age, although onset has been described as early as age 2 and as late as age 80. The onset is insidious, with subtle changes in coordination and movement developing gradually. In typical adult-onset cases, the disease may commence with increased restlessness, inability to sit still, mild but abnormal postural changes, and quick jerk-like or flicker-like movements of the fingers, limbs, and trunk (chorea). In addition to these abnormal movements, there are abnormalities of eye movement and fine motor coordination. These are manifest by visual distractibility and difficulty maintaining eye contact. Individuals may be noted by others to move their head prior to or concomitant with a change in the direction of gaze. A person may drop objects frequently and have difficulty with tasks such as buttoning, sewing, playing a musical instrument, or writing.

In conjunction with or even prior to the development of these motor abnormalities, individuals may display behavioral or cognitive difficulties. Mood is commonly affected and depression is the most frequent psychiatric problem. Individuals frequently display increased irritability, impulsive

113

behavior, and occasionally frankly inappropriate behavior. A small percentage of persons with HD develop psychosis with hallucinations, delusions, and paranoia.

As the disease progresses, the movement disorder worsens gradually but relentlessly (Figure 4.1). Abnormal choreic movements become more

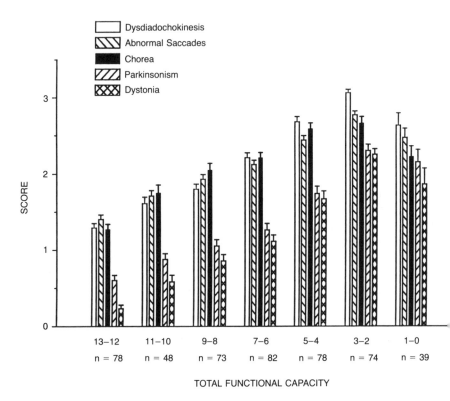

Figure 4.1: Progression of motor symptoms in HD. Cases from Venezuela who were followed over 8 years with quantitative examinations and assessments of functional capacity. Functional capacities ranged from normal (13) to totally disabled (0) and motor abnormalities ranged from 0 to 3 (0 = normal; 1 = mild; 2 = moderate; 3 = marked). Chorea, dysdiadochokinesis (abnormalities of fine motor coordination), and abnormal saccades were all detectable early in the illness. These signs worsened with progressing disease until the final stages, when they plateaued. Dystonia and parkinsonism were less prominent early in the course of the disease and then progressed. Values represent the mean q SEM of n cases in each group. *Reprinted by permission from Penney JB, Young AB, Shoulson I, et al. Huntington's disease in Venezuela: eight years of follow-up on symptomatic and asymptomatic individuals. Mov Dis 1990;5:93–99.*

frequent and larger in amplitude. Abnormal posturing (dystonia) and an unsteady, seemingly drunken gait develop. Affected persons have difficulty maintaining normal employment, and as the symptoms worsen they become unable to carry out their domestic responsibilities and eventually their activities of daily living. Over 8–15 years the patients become completely disabled and must rely on total care. Even early in the disease, speech becomes irregular and eventually difficult to understand.

By the middle to late stages of the illness, individuals have prominent dysarthria and dysphagia. Patients frequently develop aspiration pneumonia in the later stages. Although abnormalities in tone are uncommon early in the course of the disease, rigidity and dystonia become more prominent as the disease progresses. Eventually the choreic movements may actually diminish, whereas the dystonia and rigidity may become the most obvious features. Reflexes are often hyperactive even early in the disease and clonus may develop. Extensor plantar responses, however, are uncommon until the latest stages of the illness. Sensory abnormalities are infrequent and other diseases should be considered should sensory signs become obvious.

Behavior also changes as the disease progresses. Depression and apathy are common. Individuals become more withdrawn and disinterested in participating in day-to-day activities. Often this behavioral withdrawal is out of proportion to the physical disability. Problems with organization and memory also develop and these make it difficult to carry out complex tasks or even to engage in social activities. Unlike the memory disorder in Alzheimer's disease, however, individuals maintain their ability to recognize family members and friends, retain their understanding of the basic use of objects and are able to respond appropriately to emotionally charged events.

Weight loss is a striking feature of HD and it can occur prior to the onset of an overt movement disorder. Metabolic studies indicate that patients often have to consume 4,000 to 6,000 calories a day to maintain their weight. Endocrine abnormalities do not appear to account for this increased metabolic demand.

About 10% of all cases begin under the age of 20 years (so-called juvenile onset). In these individuals the motor disorder may be quite different from that observed in adults. Chorea may not develop, but instead the individual may display rigidity, tremor, and dystonia. These signs may complicate the diagnosis and therefore detailed family history is very important. Seizures and cerebellar ataxia may also manifest themselves in childhood-onset disease. Juvenile-onset cases have a more rapid and relentless course than the typical adult-onset case.

The average duration from diagnosis to death is 15 years in adult-onset cases and 8–10 years in juvenile-onset cases. Many individuals, of course, live longer depending on the quality of their long-term care and their ability to resist infection and other diseases.

MENDELIAN HERITABILITY

HD is inherited in a strictly autosomal dominant Mendelian fashion. Males and females are affected equally. Fifty percent of offspring of affected individuals will develop the disease. Although the age of onset is variable, all individuals inheriting the gene will eventually display signs of the illness if they live long enough. Studies of homozygotes for the *HD* gene reveal no gene dosage effect.

Those individuals who inherit the illness from their father have a statistical chance of developing the disease several years earlier than those who have inherited it from their mother (Figure 4.2). Approximately 80% of individuals developing the illness below the age of 15 have inherited the disease from their father. The mutation rate for this disease appears to be either nonexistent or very low. All phenotypic variations in the disease are seen within single large families. Twin studies suggest that the age of onset is genetically determined, since there is excellent concordance of age of onset in identical twins. In contrast, there can be considerable phenotypic variation between identical twins.

Most epidemiologic studies suggest that the gene originated in northern Europe, since its prevalence is highest in these regions and in areas of the world that had frequent trade with northern Europe. The prevalence of the disease is lower in Asia and in central Africa. A very high prevalence of HD exists along the shores of Lake Maracaibo in Venezuela due to the presence of a large family with the disease in this region.

MOLECULAR GENETICS

Localization of Genetic Defect

The autosomal dominant inheritance, high penetrance, and relatively straightforward diagnosis of HD made this disorder an ideal test case for application in the early 1980s of what was then a completely new strategy for approaching genetic disease. The success achieved in the HD model has since been repeated in numerous other inherited disorders and has become the method of choice for attacking genetic defects for which no protein defect has been identified.

The approach is based on a merging of the rules of Mendelian genetics with the observed high rate of sequence variation in human DNA. DNA sequence differences, most easily monitored as alterations in the pattern of cutting of human DNA by restriction enzymes and hence

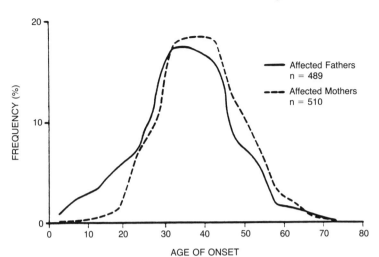

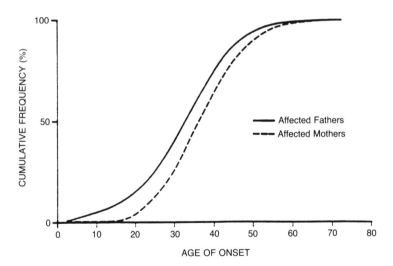

Figure 4.2: Age of onset distribution for offspring of affected fathers versus affected mothers (top) and cumulative frequency of age of onset in offspring of affected fathers versus affected mothers (bottom). *Reprinted by permission from Conneally PM. Huntington's disease: genetics and epidemiology. Am J Hum Genet 1984;36:506–526.*

termed restriction fragment length polymorphisms (RFLPs), provide an abundant supply of genetic markers that can be used to trace the inheritance of the chromosomal region in which they lie. If the genetic defect in a disease pedigree is located in the same chromosomal region as a given genetic marker (a particular variant in the same chromosomal region), a specific variant of that marker will display a correlated pattern of inheritance with the disease. The strength of the correlation, or conversely the percentage of cases in which the expected pattern of the marker is not transmitted with the disease, provides a measure of the probable distance between the marker and the defect. The cases in which the marker does not segregate with the gene result from meiotic recombination, in which material is exchanged between the two homologous chromosomes during gamete formation. On average, a frequency of recombination of 1% (or a correlation of 99%) will be obtained if two genes are 10^6 bp apart. Thus, for determining the chromosomal location of any disease gene, the inheritance of genetic markers drawn from throughout the genome must be traced in the families with the disorder. When one is found that displays a correlated pattern of inheritance with the disease and is therefore said to be "genetically linked" to the disorder, it is appropriate to conclude that the genetic defect is somewhere in the vicinity of the marker used.

The chromosomal localization of a given disease gene by this approach has two major consequences. In the immediate term, linked markers can be used to improve diagnostic capacity by permitting prenatal or presymptomatic testing for inheritance of the defect in appropriate families. The more profound impact, however, is the avenue that the discovery of a linked marker opens to isolation and characterization of the primary cause of the disorder, the disease gene itself, based not on a knowledge of the protein it encodes but only on its position on a chromosome.

In 1983, the defect causing HD was the first disease gene mapped using only this strategy. It was assigned to chromosome 4 by the discovery of G8, a marker later designated *D4S10,* that showed a very high level of coinheritance with the disorder in a large Venezuelan HD family containing dozens of affected members. This linkage was confirmed in several other populations, and the accumulated data from families throughout the world indicated that this is not a genetically heterogeneous disorder in that all cases of HD probably result from mutations in the same gene.

Fine Structure of the *HD* Region

The position of the *HD* gene was further delineated by combining the results of several different mapping approaches. Deletion mapping and dosage studies using the G8 probe in patients with the congenital anomaly Wolf-Hirschhorn syndrome, caused by heterozygous deletion on the short arm of chromosome 4, placed the marker in the terminal 4p16 cytogenetic band. In situ hybridization with the probe confirmed this result and suggested a location for the marker in the distal half of 4p16. The construction of panels of somatic cell hybrids with chromosome 4 deletions or translocations suggested that the marker was located in the proximal half of 4p16.3, the terminal cytogenetic sub-band. Finally, the analysis of meiotic recombination events in HD pedigrees suggested that the genetic defect was located 3–4% recombination away from G8 in the direction of the telomere. Thus, the disease gene was confined to a cytogenetic region expected to contain at most 7×10^6 bp of DNA, bordered on one side by the G8 marker and on the other by the end of the chromosome arm (Figure 4.3).

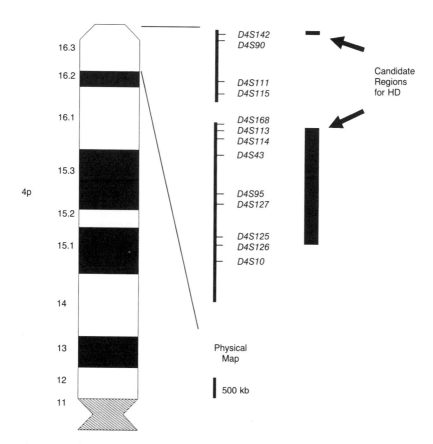

Figure 4.3: Schematic diagram of chromosome 4p showing the order of DNA markers on the physical map of 4p16.3 and the size and position of the two candidate regions in which the gene defect might lie.

To accelerate the search for the disease gene within this region, a multigroup collaborative effect catalyzed by the Hereditary Disease Foundation was formed (J. Gusella, MGH; D. Housman, MIT; Hans Lehrach, ICRF; John Wasmuth, UC Irvine; Francis Collins, Michigan; Peter Harper, Cardiff). This cooperation has produced highly detailed physical and genetic maps of the 4p16.3 sub-band but has not yet accomplished the difficult task of identifying the HD defect among the dozens of genes located in the area.

The physical mapping, using a combination of somatic cell hybrids and pulsed field gel electrophoresis (PFGE), indicates that the segment of the chromosome from G8 to the telomere contains about 6×10^6 bp of DNA, of which perhaps 35% has now been cloned. The genetic map spans about 6% recombination from $D4S10$ to $D4S90$, a marker within 300,000 bp of the telomere. However, the recombination events are not distributed equally across the physical map, with about half of the genetic distance corresponding to the 5–10% of the map adjacent to G8 and the other half comprising the remaining several million base pairs to the telomere.

The placement of the HD gene relative to the markers on the physical and genetic maps (Figure 4.3) and the corresponding sites on the physical map has been hampered by contradictory clues provided by recombination events in HD families. Several predict that the HD gene is located in the last 100,000 bp of DNA on the chromosome. Consequently, the telomere and this entire candidate region have been isolated as an artificial chromosome in yeast. However, other recombination events predict an internal 4p16.3 location for HD that spans some 2×10^6 bp of DNA. Alleles at certain polymorphic sites within this internal candidate region show association with HD, in that they are present more frequently on HD chromosomes than would be expected from their frequency in the normal population. Unfortunately, to date, these associations (termed "linkage disequilibrium") have not significantly narrowed the size of this candidate region.

Clues from the Expression of the HD Gene

The final step of isolating the HD gene has not yet been achieved, and has proven to be a difficult task due to both the uncertainties in precisely locating the candidate sequences and to the characteristics of the disease itself. The genetic linkage strategy has been applied in the past decade to many other inherited disorders, and in a few of these, chromosomal localization of the genetic defect has already led to its isolation. In most cases, the identification of the disease gene was facilitated by the availability of translocations or deletions that directly interrupted the gene, thereby causing the disorder. Only in the case of cystic fibrosis was such a rearrangement of the disease gene unavailable. In this case, however, the likelihood that the genetic defect lay in a protein involved in ion transport had already been established. In all cases, however, the gene targeted for isolation was one in which the mutations leading to a clinical phenotype were inactivating mutations.

Huntington's disease stands in stark contrast to these other disorders. Unlike almost every other human disorder so far characterized, the Huntington's disease gene causes a completely dominant phenotype. Individuals who possess two copies of the mutant *HD* gene (*HD* homozygotes) display no significant difference in age of onset, clinical manifestation, or rate of progression of the disorder relative to the typical HD victim who has one mutant and one normal copy of the gene. Thus, the HD mutation cannot result in the inactivation of an essential protein or homozygosity would be lethal. Similarly, these observations imply that the normal homologue of the mutant *HD* gene present in most HD victims does not have any effect on slowing the disease process. It is likely, therefore, that unlike most other genetic disorders characterized at the molecular level, Huntington's disease is due to a "gain of function" mutation rather than to an inactivating one. This possibility is consistent with the absence of HD-like symptoms in patients known to have chromosomal rearrangements within the *HD* candidate region, and the absence of any detected structural abnormalities on HD chromosomes.

The search for a "gain of function" mutation, which could be as subtle as a single base change in an otherwise normal gene, may require detailed sequence and expression of a large number of genes unless the candidate region can be further narrowed. In addition to comparing directly the sequence of candidate transcripts from normal and HD chromosome 4s, analysis of levels of mRNA could suggest either overexpression or inappropriate expression of the *HD* transcript, analogous to known "gain of function" mutations in *Drosophila*. While there is no guarantee that the normal site of expression of the *HD* transcript parallels the regions of neuronal loss in the disease, there is no consistent evidence for an effect of the *HD* mutation in peripheral tissues. Thus, analysis of expression of candidate sequences in HD and normal brain appears to be the most likely route for identifying meaningful quantitative changes in expression.

Identification of the *HD* Gene

Any sequence alteration in an *HD*-derived transcript could represent the mutation, but could also simply be due to polymorphic variation. The observed linkage disequilibrium suggests that a limited number of independent mutations, possibly only one, are represented in the majority of HD patients and that any rare polymorphism present near the disease gene when the mutation occurred will still show strong association with the disorder. Thus, proof that a particular candidate gene is the *HD* gene may require demonstrating the sequence alteration caused by a new mutation. The observed rate of new mutation in HD is extremely low, with no unequivocally proven case. However, several possible new mutants exist in which such sequence alterations can be sought once the number of candidate genes is reduced.

Final proof of the identity of the *HD* candidate may require reproducing the neuropathology of the disorder in an animal model. The dominant nature of the mutation suggests that this might be possible using transgenic mouse technology. A major piece of information that will affect the potential success of this approach is the determination of when during life the *HD* gene is expressed. Although clinical and neuropathological signs of the disorder are not detectable until adult life, it is not certain whether the *HD* gene is expressed from early in development or only begins to be expressed in middle age. If the effect of the gene is to cause abnormal development of the brain that later leads to the characteristic neuropathology, such changes might be detectable in a mouse transgenic model. If, however, the disease gene only switches on late in life, the lifespan of the mouse may be too short for any negative effects of the gene to be seen.

Despite the difficulty of distinguishing the *HD* gene from its many neighbors, the finite size of the candidate region and the increasing sophistication of molecular genetic analysis indicate that finding the genetic defect is only a matter of time. Once the *HD* gene has been isolated, a detailed characterization of its structure, sequence, and pattern of expression should provide clues to the function of the normal gene and the manner in which the HD mutation causes abnormal function. It is to be hoped that this knowledge will also suggest targets for the development of rational therapies that will slow or even halt the onset or progression of this devastating disorder.

MOLECULAR MEDICINE

Diagnostics

Prior to the identification of markers close to the *HD* gene, the diagnosis of this disorder was based purely on clinical examination, family history, and the exclusion of several rare metabolic disorders that can also cause progressive chorea. In symptomatic individuals with a family history, the diagnosis is fairly straightforward and is based on the presence of the characteristic motor features described previously and concomitant cognitive dysfunction. Often, however, family histories are inconclusive or nonexistent. In these situations only a tentative diagnosis can be rendered, based on the exclusion of other illnesses, the presence of progressive atrophy of the caudate nuclei and putamen, and ultimately the neuro-pathological abnormalities.

A major dilemma in this disease has been the inability to predict a diagnosis for persons at risk for HD. No simple and unambiguous clinical or laboratory test is available to identify which at-risk individuals have inherited the gene. Recent data using positron emission tomography would suggest that cerebral caudate glucose metabolism becomes abnormal several years prior to the onset of clinical symptoms in many cases. These findings, however, have not yet been universally observed and some individuals have been shown to develop hypometabolism only after or in conjunction with the earliest clinical symptoms. It is now possible to use DNA markers closely linked to the *HD* gene as a diagnostic test in certain at-risk individuals. This test utilizes the capability of doing linkage analysis in families with known HD and comparing the marker status of the known affected parent, other affected family members, escapees, and the unaffected parent with the marker status of the at-risk individual being tested. Usually at least two affected individuals who are close relatives of the at-risk person being tested must donate their blood for linkage analysis. The test is approximately 98% accurate when the pedigree and marker status of the individuals are optimal. The accuracy, however, can be considerably less when key individuals in the family tree are unavailable. Because this test relies on linkage analysis, it cannot be applied accurately to individuals in families where the diagnosis of HD is questionable.

The presymptomatic linkage-analysis test for HD has been available now for over five years. Numerous individuals have undergone presymptomatic testing as part of a detailed structured clinical protocol. Overall, the number of individuals seeking testing is lower than initially predicted and a higher percentage of these individuals than expected on statistical grounds have been found not to carry the *HD* gene once the linkage testing has been completed. Presumably, some selection process is occurring in the individuals seeking testing. Depression is common after receiving a positive presymptomatic test and these individuals must be followed closely with psychological counseling and support.

The diagnostic linkage testing can be applied as a prenatal test. The prenatal test requires fewer participating family members than are required for testing the asymptomatic at-risk individual. With a limited pedigree that includes the at-risk individual, the spouse of the at-risk individual, and the affected and unaffected parents of the at-risk individual, it is possible to predict whether the fetus is either likely to be free of carrying the *HD* gene or to have the same risk as the at-risk parent. This so-called "nondisclosing" test allows an at-risk individual to identify those embryos that have a very low risk of having the *HD* gene and those that may or may not carry the gene but have the same risk as the at-risk individual him/herself.

Using this method of testing, the at-risk individual whose embryo is being tested is not diagnosed during the process. Fewer than expected individuals have participated in this form of testing presumably due to the reluctance in aborting a fetus that only has a 50% chance of developing HD.

Obviously when the gene itself is identified, the diagnostic process will be vastly simplified. It will then be possible to diagnose people directly with or without a family history of the disease and it will be possible to get very accurate prenatal testing.

Prevention

Although the use of linkage-analysis presymptomatic and prenatal testing affords accurate methodologies for the prevention of the disease, few people overall have participated in these studies. Without the availability of any effective therapy, the knowledge of carrying the gene for a certainly lethal disease is psychologically overwhelming. Likewise, the biological drive for having children has overridden many at-risk individuals' concerns of developing the disease and passing it on to their children. Reproductive rates in affected individuals are equal to those in the general population. Even in individuals participating in careful genetic counseling programs prior to reproduction, the reproductive rates are only slightly below those of normal individuals. When the *HD* gene is identified, at-risk individuals may be more willing to use this technology to prevent passing this gene on to offspring.

Pathophysiology

The consequence of inheriting the *HD* gene is the gradual loss of neurons, predominantly in the caudate nucleus and putamen of the brain. These two regions lie in the central portions of the brain and are extremely important in the control of coordinated movement. Other nerve cells in the cerebral cortex, thalamus, and cerebellum may also be affected in those with the disease, but it is not certain whether this loss is a primary consequence of the disease or a secondary process. Within the caudate nucleus and putamen, there is an ordered loss of neurons as the disease progresses. It begins in the medial dorsal portions of the caudate nucleus and the putamen. With advancing illness, the involvement progresses ventrolaterally.

Neuropathological studies suggest that not all nerve cells within the caudate nucleus are affected equally in this disease. Certain subsets of nerve cells are relatively spared and others appear affected very early in the illness. In particular, interneurons using somatostatin and neuropeptide Y as neurotransmitters, and interneurons using acetylcholine as a neurotransmitter appear to be relatively unaffected by the illness. In contrast, projection neurons using the neurotransmitter GABA appear to be affected relatively early (Figure 4.4). In fact, subsets of these GABA

NORMAL

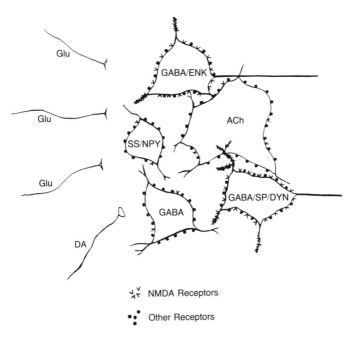

Glu

Glu

Glu

DA

GABA/ENK

ACh

SS/NPY

GABA/SP/DYN

GABA

NMDA Receptors

Other Receptors

HUNTINGTON'S DISEASE

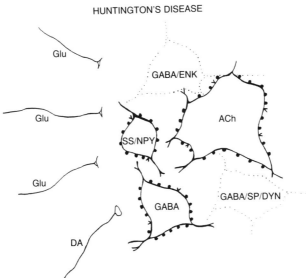

Glu

Glu

Glu

DA

GABA/ENK

ACh

SS/NPY

GABA/SP/DYN

GABA

Figure 4.4: Schematic diagram of the neuronal types in the normal (top) and HD (bottom) caudate/putamen. Glutamate (Glu) afferents from cerebral cortex project to caudate/putamen. Dopamine (DA) afferents emanate from substantia nigra pars

projection neurons may be differentially affected in the illness. In studies of presymptomatic individuals who were gene carriers by linkage-analysis testing, GABA neurons projecting to the lateral globus pallidus and substantia nigra pars reticulata were found to be affected very early in the disease. In these individuals there were also abnormalities in the N-methyl-D-aspartate subtype of excitatory amino acid receptor.

One popular theory about the pathophysiology of *HD* is that there is either an excess of endogenous neurotoxic substances in this illness or that the cells in the caudate nucleus become excessively sensitive to the brain's own natural toxic chemical messengers. Future neurobiologic research should address the issues of which cells are affected earliest in the disease, what are the special properties of these neurons, whether glial factors or endogenous toxins are involved, and whether there are steps at which the degenerative process can be halted. The identification of the *HD* gene should provide important clues as to the pathophysiology of the disease.

Therapeutic Options and Directions

Until the gene is identified, therapeutic strategies are merely speculative. Based on the excitotoxic hypothesis, it is conceivable that antagonists of excitotoxins might delay the onset and progression of the illness. Therapeutic trials using these agents have been proposed. Based on the observation that HD is a true dominant disorder and that homozygotes for the disease are no different than heterozygotes, it is unlikely that gene therapy will prove effective in *HD*.

SELECTED REFERENCES

Albin RL, Young AB, Penney JB. The functional anatomy of basal ganglia disorders. Trends Neurosci 1989;12:366–375.

Bates GP, MacDonald ME, Baxendale S, et al. Defined physical limits of the Huntington disease gene candidate region. Am J Hum Genet 1991;49:7–16.

Figure 4.4, continued. compacta. The majority of caudate/putamen neurons are GABAergic and most (but not all) of these cells are medium spiny neurons that project to areas outside the caudate/putamen. One major subclass of these neurons co-contain enkephalin (ENK) and GABA; another class co-contains GABA, substance P (SP), and dynorphin (DYN). These neurons are affected relatively early in HD and it has been hypothesized that they may have a relatively higher concentration of N-methyl-D-aspartate (NMDA) on their surface than do the interneurons. Two main classes of interneurons, the small somatostatin/neuropeptide Y (SS/NPY) interneurons and large acetylcholine (ACh) neurons, are relatively spared in HD.

Conneally PM. Huntington's disease: genetics and epidemiology. Am J Hum Genet 1984;36:506–526.

Conneally PM, Haines JL, Tanzi RE, et al. Huntington disease: no evidence for locus heterogeneity. Genomics 1989;5:304–308.

Folstein SE. Huntington's disease. A disorder of families. Baltimore: The Johns Hopkins University Press, 1989.

Gusella JF. DNA polymorphism and human disease. Ann Rev Biochem 1986;55:831–854.

Gusella JF. Location cloning strategy for characterizing genetic defects in Huntington's disease and Alzheimer's disease. FASEB J 1989;3:2036–2041.

Gusella JF, Tanzi RE, Anderson MA, et al. DNA markers for nervous system diseases. Science 1984;225:1320–1326.

Gusella JF, Wexler NS, Conneally PM, et al. A polymorphic DNA marker genetically linked to Huntington's disease. Nature 1983;306:234–238.

Hayden MR. Huntington's chorea. Berlin: Springer-Verlag, 1981.

Huntington G. On chorea. Med Surg Rep 1872;26:320–321.

MacDonald ME, Haines, JL, Zimmer M, et al. Recombination events suggest possible locations for the Huntington's disease gene. Neuron 1989;3:183–190.

MacDonald ME, Lin C, Srinidhi L, et al. Complex patterns of linkage disequilibrium in the Huntington disease region. Am J Hum Genet 1991;49:723–734.

Martin JB, Gusella JF. Huntington's disease. Pathogenesis and management. N Engl J Med 1986;315:1267–1276.

Mazziotta JC, Phelps ME, Puhl JJ, et al. Reduced cerebral glucose metabolism in asymptomatic subjects at risk for Huntington's disease. N Engl J Med 1987;316:357–362.

Meissen GJ, Myers RH, Mastromauro CA, et al. Predictive testing for Huntington's disease with use of a linked DNA marker. N Engl J Med 1988;318:535–542.

Penney JB, Young AB, Shoulson I, et al. Huntington's disease in Venezuela: eight years of follow-up on symptomatic and asymptomatic individuals. Mov Dis 1990;5:93–99.

Shoulson I. Huntington's disease. In: Asbury AK, McKhann GM, McDonald WI, eds. Diseases of the nervous system. Clinical neurobiology. Philadelphia: WB Saunders, 1986;1258–1267.

Vonsattel JP, Myers RH, Stevens TJ, et al. Neuropathological classification of Huntington's disease. J Neuropathol Exp Neurol 1985;44:549–557.

Wexler NS, Young AB, Tanzi RE, et al. Homozygotes for Huntington's disease. Nature 1987;326:194–197.

Young AB, Penney JB, Starosta-Rubenstein S, et al. Normal glucose metabolism in persons at risk for Huntington's disease. Arch Neurol 1987;44:254–257.

CHAPTER 5

HPRT Deficiency: Lesch–Nyhan Syndrome and Gouty Arthritis

C. Thomas Caskey
Belinda J. F. Rossiter

HYPOXANTHINE GUANINE PHOSPHORIBOSYLTRANSFERASE

Hypoxanthine guanine phosphoribosyltransferase (HPRT; inosine mono-phosphate: pyrophosphatase phosphoribosyltransferase; EC 2.4.2.8) is one of the enzymes responsible for the salvage of preformed purine and pyrimidine bases present in the mammalian cell as the result of normal nucleic acid turnover (Figure 5.1). HPRT catalyzes the transfer of phos-phoribose from 5'-phosphoribosyl-1-pyrophosphate to the 9 position of hypoxanthine or guanine yielding 5'-inosine monophosphate or 5'-guanine monophosphate, respectively. Mammalian cells are also capable of making these nucleotides de novo by alternative pathways.

The native HPRT protein is normally a tetramer of identical 217–amino acid subunits. HPRT is a soluble cytoplasmic enzyme found in all human tissues at a low level and at higher levels in the brain, particularly the basal ganglia, suggesting that there is a greater dependence on salvage pathways of purine synthesis rather than de novo synthesis in this region.

CLINICAL FEATURES OF HPRT-DEFICIENCY DISORDERS

The Lesch-Nyhan Syndrome

In 1967, Seegmiller et al. found that a deficiency in HPRT activity was the primary lesion in the cells of patients suffering from Lesch–Nyhan syn-drome (LN), a devastating disorder characterized by hyperuricemia, choreo-

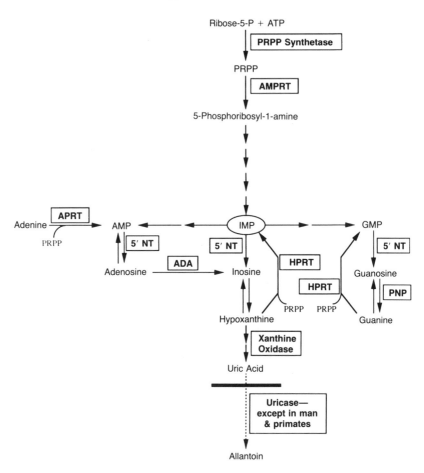

Figure 5.1: Simplified diagram of purine synthesis. ADA = adenosine deaminase; AMPRT = amidophosphoribosyltransferase; APRT = adenine phosphoribosyltransferase; GMP = 5'-guanine monophosphate; IMP = 5'-inosine monophosphate; 5' NT = 5' nucleotidase; PNP = purine nucleoside phosphorylase; PRPP = 5'-phosphoribosyl-1-pyrophosphate.

athetoid movements, spasticity, hyperreflexia, mental retardation, and compulsive self-mutilating behavior. The *HPRT* gene was localized to the X chromosome in man as the result of pedigree analysis of families with LN; somatic cell hybridization studies further defined the location as the Xq26 region. The inheritance of LN follows a classic recessive X-linked pattern, illustrated in Figure 5.2. Estimates of the frequency of LN in the population have ranged from 1 in 380,000 live births to 1 in 10,000 male births.

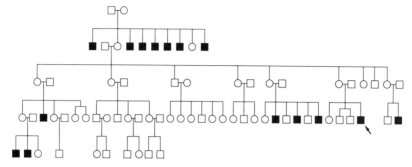

Figure 5.2: Pedigree of the family of LN patient D.B., showing X-linked recessive inheritance. Circle represents female; square represents male; filled square represents affected male; arrow indicates patient D.B. *Redrawn with permission from Nyhan WL, Pesek J, Sweetman L, Carpenter DG, Carter CH. Genetics of an X-linked disorder of uric acid metabolism and cerebral function. Pediatr Res 1967;1:5–13.*

Symptoms of HPRT deficiency are not usually apparent in carrier heterozygote females.

Patients with LN are clinically normal at birth but developmental delay is evident after about six months, and choreiform movements appear within the first year. The most intriguing and miserable symptom, that of self-mutilation, may begin as early as the first year or as late as 16 years. Patients begin by biting their lips or buccal mucosa and frequently progress to the biting of fingers and hands. The destructive urge is often so great that arm restraints or even dental extraction may be necessary in order to prevent serious injury. Sufferers of LN may show some improvement in aggressive and self-mutilating behavior as they grow older, probably as the result of greater self-discipline, but commonly die before the age of 30 from infection or renal failure.

Gouty Arthritis

Partial HPRT deficiency is associated with increased de novo purine synthesis and hyperuricemia, which results in nephrolithiasis and gouty arthritis. Approximately 20% of patients with partial HPRT deficiency exhibit spasticity, cerebellar ataxia, and mild mental retardation, but they do not become self-mutilating. It is interesting to note that although LN patients have hyperuricemia, they do not usually develop gouty arthritis, whereas most if not all patients with partial HPRT deficiency will eventually develop gouty arthritis. This may simply reflect the age of onset, however, since LN patients are not long-lived and gouty arthritis symptoms take years to become apparent.

THE *HPRT* GENE

Sequence Features

Sequencing of 57 kilobases (kb) around and including the human *HPRT* locus employed a prototype fluorescent deoxyribonucleic acid (DNA) sequencing device. Robotic equipment to assist in the template preparation and sequencing reactions, and software for the management of the sequence data were also developed during the course of the sequencing project. The coding region of the human *HPRT* gene is divided into nine exons and is dispersed over 39.8 kb DNA. The positions of the exons and details of other sequence features of the human *HPRT* gene are shown in Figure 5.3.

The human *HPRT* gene lacks consensus CAAT and TATA sequences immediately 5′ to the initiation site. The promoter region 5′ to exon 1 contains five copies of the 5′-GGCGGG-3′ sequence, a potential binding site for the Sp1 transcription factor and a pattern also found in several viral promoters, including those of the simian virus 40 early and late genes and the herpes virus thymidine kinase gene. A promoter region containing GC-rich control elements and lacking CAAT and TATA sequences is a feature of ubiquitously expressed housekeeping genes, such as phosphoglycerate kinase, adenosine deaminase, adenosine phosphoribosyltransferase, lysosomal acid phosphatase, and 5-lipoxygenase. The GC-rich region at the 5′ end of the *HPRT* gene also extends 3′ to exon 1, where there are a further two "GC boxes."

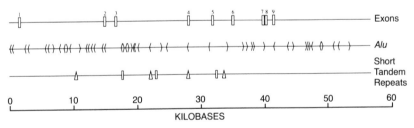

Figure 5.3: Sequence features of the human *HPRT* gene. The locations of the nine coding exons of the *HPRT* gene are shown, together with the positions of Alu repeats (and their different orientations) and two classes of short tandem repeats found within the locus. *Reprinted by permission from Rossiter BJF, Caskey CT. HPRT mutation and the Lesch-Nyhan syndrome. In: Brosius J, Fremeau RT, eds. Molecular genetic approaches to neuropsychiatric disease. San Diego: Academic Press, Inc., 1991: 97–124.*

The human *HPRT* gene contains 49 members of the interspersed middle repetitive element *Alu,* intriguingly oriented asymmetrically. The *Alu* repeats in the 5′ region of the gene are predominantly in the reverse orientation with respect to the direction of *HPRT* transcription, and those in the 3′ region of the gene are oriented randomly. In addition to the 49 *Alu* repeats, the *HPRT* gene contains a number of other repeat elements; in all, the various repetitive sequences make up 30% of the locus.

Polymorphisms

Examination of the *HPRT* sequence has revealed the presence of short tandem repeats (STRs), i.e., multiple units of dimers, trimers, and tetramers, one of which has proved to be polymorphic with seven alleles (Figure 5.4). Polymerase chain reaction (PCR) amplification of the region containing this STR results in fragments of various sizes according to the number of repeated units. Previously, there had been reports of only one restriction fragment length polymorphism (*Bam*HI) within the human *HPRT* gene with a low predicted heterozygote frequency (PHF) of 29%. Since this particular polymorphism is not very informative and the required Southern analysis is slow and tedious, the discovery of a more easily obtained and highly polymorphic (70% PHF) marker at this locus has assisted greatly in the assessment of families where one member is affected by an *HPRT* mutation.

GENE DEFECTS RESULTING IN DISEASE

Lesch-Nyhan syndrome is a genetic lethal disorder because males with LN do not reproduce. Haldane's hypothesis predicts that in instances such as this, one-third of the mutations will be new; this is approximately the case for LN. An example of a pedigree showing the occurrence of a new mutation is shown in Figure 5.5. The mutations found within the *HPRT* locus are heterogeneous and there are rarely reports of two unrelated patients with the same molecular defect. These observations, in addition to the fact that 85% of mutations in the *HPRT* genes of LN patients are undetectable by Southern analysis, present certain challenges in the efficient detection of mutations in this relatively large (40 kb) gene.

Figure 5.6 shows the in vivo *HPRT* mutations causing LN or gouty arthritis that have been described in the literature and indicates the heterogeneous nature of the defects observed, the bulk of which are point mutations and small rearrangements. It is nevertheless possible to group the point mutations in clusters, according to predicted domains in the protein structure. A mutation within the *HPRT* protein resulting in disease clearly

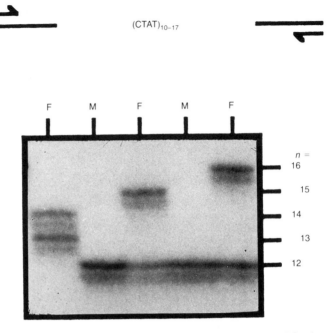

Figure 5.4: STR polymorphism in the *HPRT* gene. PCR amplification using primers flanking the polymorphic $(CTAT)_n$ sequence within the *HPRT* gene results in differently sized fragments according to the number of repeated units. α-^{32}P-deoxycytidine is included in the reaction mixture and the product is run out on a DNA sequencing gel. Following autoradiography, the banding pattern can be observed; each allele appears as a doublet as both strands are labeled. Five of seven alleles are illustrated. M, male; F, female; *n*, number of repeated units. *Reprinted by permission from Rossiter BJF, Caskey CT. HPRT mutation and the Lesch-Nyhan syndrome. In: Brosius J, Fremeau RT, eds. Molecular genetic approaches to neuropsychiatric disease. San Diego: Academic Press, Inc., 1991:97–124.*

has reduced or eliminated activity, and the location of such a mutation may tell us something about the importance of that particular region for enzyme activity or stability. It is also possible to surmise important regions of the protein by comparison of the mammalian enzyme with that of *Plasmodium falciparum*, which is less conserved, the implication being that the conserved regions are those important for structure or function.

In most cases, a mutation within the *HPRT* gene results in LN rather than gouty arthritis; the only cases of gout result from point mutations affecting single amino acids, and one deletion that removes the start codon, presumably resulting in initiation of translation in-frame further downstream.

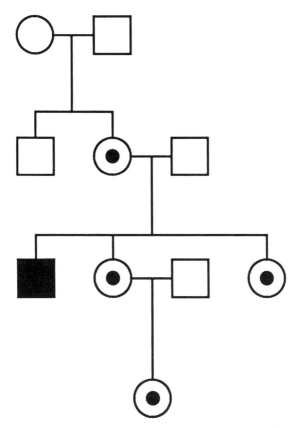

Figure 5.5: Pedigree of GM1662 family showing occurrence of new mutation. Circle represents female; square represents male; filled square represents affected male; dot in circle represents carrier female. *Redrawn with permission from Yang TP, Patel PI, Chinault AC, et al. Molecular evidence for new mutation at the HPRT locus in Lesch-Nyhan patients. Nature 1984;310:412–414. Copyright © 1984 Macmillan Magazines Ltd.*

A very detailed study of different regions of the *HPRT* protein is not likely to result solely from a collection of gouty arthritis and LN mutations because of their limited number. Such information is more likely to come from the study of mutations spontaneously occurring or induced in cultured cell lines because of the greater number available. One advantage of the in vivo system, however, is the observation of different phenotypes (gouty arthritis or LN) according to the severity of the mutation, possibly indicating more or less important roles for the affected region of the protein.

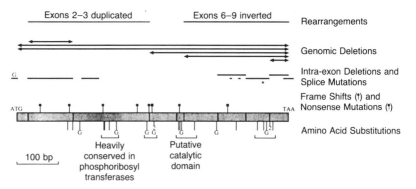

Figure 5.6: Summary of published mutations within the human *HPRT* gene. The information is taken from Rossiter et al. (1991) and only includes mutations resulting in LN or gouty arthritis; no in vitro cell mutation data are included. Mutations marked with a "G" result in gouty arthritis; all others lead to LN. The large shaded bar represents the coding region of the *HPRT* gene, divided into exons by the vertical lines. Point mutations resulting in amino acid substitutions are shown below the coding region; frameshift (1- or 2-nucleotide deletions or insertions) and nonsense mutations are shown above. The amino acid substitutions marked with "2" indicate that each of these have been observed in two unrelated individuals. The square bracketed regions correspond to domains conserved between the mammalian and *Plasmodium falciparum* HPRT proteins. Intra-exon deletions are defined as those contained within one exon; splice-site mutations usually result in the removal of an entire exon from the mRNA. The mutation marked by "*" could be a splice-site mutation or a genomic deletion. Genomic deletions are defined as those including sequences flanking exons; arrows indicate that the extent of the deletion is not known. The two rearrangements shown are also associated with deletions of unknown extent.

MUTATION DETECTION

Established Mutation Detection Methods

During the 1980s, it was possible to detect major deletions and rearrangements in the *HPRT* gene by means of Southern analysis using complementary DNA (cDNA) as a probe. Since each exon can be assigned to a band on a Southern blot, it is possible to determine approximately which exons are absent in the cases of partial or complete deletions. The determination of deletion end points in this manner, however, can be imprecise when several exons are contained within one restriction fragment. In any case, only 15% of LN patients have abnormal Southern patterns indicating DNA deletions or other major rearrangements.

The presence or absence of *HPRT* messenger ribonucleic acid (mRNA) can be determined by Northern analysis, also using cDNA as a probe. Occasionally, abnormal mRNA species have been identified, suggesting a mutation affecting splicing or an insertion or deletion event. The majority (80%) of *HPRT* mutations in vivo, however, produce normal levels of normally sized mRNA, suggesting that the defect is very small and likely to be a point mutation within the coding region. In these cases, it has previously been necessary to generate and clone *HPRT* cDNA from each mutant and to sequence the molecule in order to detect the underlying mutation.

The enzyme RNase A can cleave some mismatches in duplexes of RNA from different sources. This phenomenon formed the basis of a point mutation detection assay used to detect small alterations in the *HPRT* gene, where heteroduplexes of normal and mutant RNAs were subjected to RNase A cleavage and analysis by gel electrophoresis. This method was able to detect some mismatches but suffered because of an inability to detect all alterations.

Denaturing gradient gel electrophoresis is another approach used to detect mismatches in *HPRT* heteroduplexes, in this instance between fragments of wild-type and mutant DNA. This procedure relies on the different melting profiles of homoduplexes and heteroduplexes, and therefore their mobility when subjected to electrophoresis in gradient gels. Once the mutation is localized to a particular region, the appropriate portion of DNA can be amplified using PCR and directly sequenced to determine the nature of the mutation. Only low-melting regions of the cDNA can be analyzed in this way, however, and the procedure is sensitive to the methylation state of the DNA, possibly leading to false indications of point mutations.

New Mutation Detection Methods

Knowing the sequence of the *HPRT* locus in man, it was possible to design PCR primers for amplification of the individual exons and their immediate flanking sequences. All the exons can be amplified simultaneously on different-sized fragments in a single reaction (i.e., multiplex amplification) and viewed after electrophoresis through an agarose gel (Figure 5.7). If a particular region of the gene is absent, or there is a mutation within the binding site of the priming oligonucleotide, that region fails to amplify and is not seen on the gel. It is also theoretically possible that altered-size fragments might be seen on the gel, indicative of a small deletion or insertion within an exon, but this has not yet been observed (J.S. Chamberlain, personal communication). Primers for the PCR amplification of genomic sequences must be chosen carefully to have similar melting properties, to

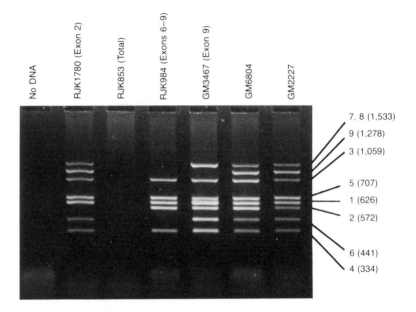

Figure 5.7: Multiplex amplification of the human *HPRT* locus. All nine exons of the *HPRT* gene were amplified on eight separate DNA fragments using 16 oligonucleotide primers in a single PCR reaction. The lane containing no DNA was a control to ensure that there was no spurious amplification due to endogenous contaminating sequences. Four examples of complete or partial *HPRT* gene deletions are shown. The cell lines GM6804 and GM2227 display the normal multiplex pattern of eight bands. The exons contained within the individual fragments and the size of the amplification units are shown on the right. *Reprinted by permission from Gibbs RA, Nguyen P-N, Edwards A, Civitello AB, Caskey CT. Multiplex DNA deletion detection and exon sequencing of the hypoxanthine phosphoribosyltransferase gene in Lesch-Nyhan families. Genomics 1990;7:235–244.*

generate fragments of different sizes so that they can be resolved by electrophoresis, and so that they do not lie within repeat regions nor hybridize to each other. It follows, therefore, that sequence should be carefully examined and interpreted before the designing of PCR oligonucleotides to avoid the waste of generating ineffective primers.

The entire coding region of the human *HPRT* gene can be amplified in one fragment when mRNA is used as a template for the synthesis of single-strand cDNA, which is then used as the substrate in a PCR reaction. The products of this reaction can then be sequenced directly, without the need for further subcloning, using manual or automated techniques. Alternatively, the products of the multiplex amplification described above can be sequenced directly, yielding the sequence not only of the coding regions, but also the flanking splice sites. The ability to amplify and sequence regions of the gene, rather than relying on obtaining sequence from mRNA, means that sequence information can be obtained even when mRNA is not produced. Such mutations have previously been refractory to analysis.

An overall strategy has therefore been developed using PCR analysis to screen for deletions in the *HPRT* gene, followed by further analysis of the PCR products by automated direct sequencing to yield information at the nucleotide level. Such a strategy can theoretically detect virtually all possible disease-causing mutations at the *HPRT* locus.

CARRIER DETECTION

Established Carrier Detection Methods

Apart from subtle abnormalities in de novo purine metabolism, *HPRT* heterozygotes in general are clinically asymptomatic. The process of random X-chromosome inactivation in a heterozygote female predicts that mosaicism will be observed, i.e., a mixture of cells expressing one or other allele, depending on which X chromosome is active; clonal populations derived from a single cell will display the same X-inactivation profile as the parent cell. If the inactivation is truly random, then 50% of the cells would be expected to express one allele and 50% would be expected to express the other. Phenotypic methods of carrier detection for HPRT deficiency have assumed that such random inactivation has taken place and that 50% of cells derived from such a female will lack HPRT activity. While it has been possible using cultured fibroblasts and hair follicles to detect HPRT-deficient cells in obligate heterozygotes, it seems that the mutant allele appears at a reduced frequency, if at all, in lymphocytes and erythrocytes. This might suggest that there is a special advantage for cells utilizing the salvage pathway of purine biosynthesis in bone marrow, and makes the task of detecting carriers of HPRT deficiency more difficult.

New Carrier Detection Methods

Molecular methods of carrier detection are not subject to the problem of nonrandom X inactivation since whether or not the mutant allele is expressed, it remains present at the expected genotypic frequency of 50% in the heterozygote female. There are therefore advantages in developing methods of detecting mutations in the *HPRT* gene in order to make carrier testing for LN more reliable. It is also desirable to increase the ease and speed at which carrier-detection assays can be performed. The methods described above of multiplex PCR amplification and direct sequencing of PCR products to detect large or small alterations in the *HPRT* gene take only a few hours to perform and are relatively straightforward. It is not only possible to detect mutations in affected individuals in this manner but also to detect a mutation present in a carrier heterozygote against the background of a normal gene, since at the sequence level it is possible to distinguish mutant and wild type even when they are superimposed on each other.

With the discovery of the highly polymorphic STR within the *HPRT* gene, it is not even necessary to determine *HPRT* sequences from family members for carrier testing to be performed. Once the "phase" of the *HPRT* mutation is discovered, i.e., the polymorphic STR allele with which the mutation is associated, then simple haplotyping can be performed to assess other family members.

APPLICATIONS TO OTHER SYSTEMS

The techniques described previously of automated sequencing, multiplex PCR deletion analysis and direct sequencing of PCR products clearly have application to many other disease loci and can also be used for the study of mutations in somatic cells. The ability to determine DNA sequence and variation rapidly and cheaply has global uses such as sequencing the entire human genome, polymorphism detection, and evolutionary studies. Highly polymorphic STR markers within the human genome can be used to great advantage in genetic mapping, medical diagnostics, paternity testing, and forensic science applications.

PATHOPHYSIOLOGY OF HPRT-DEFICIENCY DISORDERS

HPRT Deficiency in Humans

As can be deduced from Figure 5.1, a reduction in HPRT activity results in an accumulation of hypoxanthine, a substrate for xanthine oxidase. This in turn leads to an excess of uric acid, which causes the symptoms of gout. Although the manifestation of hyperuricemia is quite obvious, the specific relationship between the lack of HPRT activity and the profound neurologic symptoms of LN remains unclear. It is nevertheless intriguing that a distinct pattern of human behavior is associated with a single biochemical defect; defining the mechanism of this disturbance opens up the possibility of a greater understanding of behavior and its biochemical basis.

It has been suggested that the accumulation of toxic metabolites, derived from hypoxanthine, may cause the behavioral problems of LN. This, however, is unlikely because some patients with partial HPRT deficiency have equally high levels of hypoxanthine in the cerebrospinal fluid but lack neurological symptoms. In addition, treatment of hyperuricemia with allopurinol (which inhibits xanthine oxidase) increases hypoxanthine levels but does not have any effect on the behavioral symptoms of LN.

Dopamine pathways in the basal ganglia region of the brain are implicated in the pathogenesis of LN because (1) this region normally exhibits the highest levels of HPRT activity and low levels of de novo purine synthesis, (2) indices of dopamine function in the basal ganglia of LN patients are reduced by 70–90%, and (3) involuntary movements of LN patients are reminiscent of disorders of the extrapyramidal motor system. It has been proposed that the HPRT enzyme defect impairs central nervous system (CNS) morphogenesis or interferes with normal physiological function by causing a deficiency of purine nucleotides. The basal ganglia in particular would be particularly vulnerable with its greater dependency on de novo purine synthesis to maintain nucleic acid pools. The blockade of purine salvage appears to have its major impact on CNS function during the time of proliferation of dopamine terminals into the striatum; this occurs during the perinatal period and is also the time at which LN symptoms are first noticed.

HPRT Deficiency in Mice

Two groups have reported the generation of transgenic mice lines completely lacking HPRT activity. Contrary to expectation, these mice display no symptoms associated with their HPRT deficiency, whether hyperuricemia, gout, or behavioral. Striatal dopamine levels were depleted compared with wild-type mice, but only by 19%. Clearly there are differences between mouse and man in terms of neurological dependence on salvage purine pathways; these differences may be anatomical or biochemical. It is interesting to note that uric acid in mice is further degraded to allantoin by the enzyme uricase; the gene for this enzyme is inactive in man. Rather than increase our understanding of HPRT deficiency in man, the generation of symptomless HPRT-deficient mice has further complicated the issue!

PREVENTION AND THERAPEUTIC OPTIONS

Conventional Therapy

Hyperuricemia associated with HPRT deficiency can be controlled effectively with allopurinol, which inhibits the enzyme xanthine oxidase. This treatment reduces serum and urine uric acid levels and thus prevents uric acid crystalluria, urate nephropathy, nephrolithiasis, and gouty arthritis, but has no effect on the neurological symptoms of LN. Since treatment with allopurinol results in the excretion of large amounts of xanthine and hypoxanthine, care must be taken to titrate the dose carefully and to maintain a high urine volume, since the relative insolubility of xanthine can result in the development of xanthine stones.

In addition to allopurinol therapy, many forms of treatment have been attempted as cures for the neurological symptoms of LN. Some have been beneficial in some instances, but the findings are not consistent and no one regime has been found to be universally effective. One approach has been to supply known feedback inhibitors of de novo purine synthesis, or precursors of these compounds. Of these agents, adenine has been the most promising, but it has also led to the side-effects of 2,8-dioxyadenine nephrotoxicity. Other strategies have included exchange blood transfusion; tetrabenazine; thiopropazate; chlorpromazineorotic acid; α-methyldopa, carbidopa and levodopa; high protein and monosodium glutamate; l-5-hydroxytryptophan; and behavior modification therapy. There is a heterogeneity in response to these various treatments, and even one case of improvement when the patient elected to receive no medical treatment at all. As there is no cure for LN at this time, it is imperative that prenatal diagnosis and carrier testing be as accurate as possible.

Genetic Therapy

Since there is no prospect in the immediate future of a successful treatment for LN using conventional means and because genetic material from the *HPRT* gene is available, this devastating disorder could be considered a candidate for gene replacement therapy, since it is possible to deliver a functional *HPRT* gene to HPRT-deficient cells and to mice using retroviral vectors.

CONCLUSIONS

The recent technological advances described here have vastly increased the ability to detect rapidly and accurately new mutations in the *HPRT* gene.

While there has been no major advance in the development of an effective therapy for the devastating neurological symptoms of the Lesch-Nyhan syndrome, the procedures for carrier testing and prenatal diagnosis have improved greatly. Although these new techniques are revolutionary, there will undoubtedly be further developments and improvements in the future as new procedures become available. The use of radioisotopes is costly and a potential danger for laboratory personnel, so the continued transfer to fluorescent technology would reduce the cost and improve the safety of carrying out these procedures. The ability to execute a preliminary scan to recognize the exon containing a point mutation would mean an immediate ninefold enhancement in the efficiency of detecting *HPRT* mutations, since the gene has nine exons and it would only be necessary to sequence the affected one. These and other modifications will improve further an already vastly enhanced field of investigation.

ACKNOWLEDGMENTS

C. Thomas Caskey is an Investigator of the Howard Hughes Medical Institute, and Belinda J. F. Rossiter is the recipient of an Arthritis Foundation Fellowship.

SELECTED REFERENCES

Baumeister AA, Frye GD. The biochemical basis of the behavioral disorder in the Lesch-Nyhan syndrome. Neurosci Biobehav Rev 1985;9:169–178.

Brennand J, Konecki DS, Caskey CT. Expression of human and Chinese hamster hypoxanthine-guanine phosphoribosyltransferase cDNA recombinants in cultured Lesch-Nyhan and Chinese hamster fibroblasts. J Biol Chem 1983; 258:9593–9596.

Cariello NF, Scott JK, Kat AG, Thilly WG, Keohavong P. Resolution of a missense mutant in human genomic DNA by denaturing gradient gel electrophoresis and direct sequencing using in vitro DNA amplification: HPRT$_{Munich}$. Am J Hum Genet 1988;42:726–734.

Davidson BL, Tarlé SA, Palella TD, Kelley WN. Molecular basis of hypoxanthine-guanine phosphoribosyltransferase deficiency in ten subjects determined by direct sequencing of amplified transcripts. J Clin Invest 1989;84:342–346.

Edwards A, Voss H, Rice P, et al. Automated DNA sequencing of the human HPRT locus. Genomics 1990;6:593–608.

Finger S, Heavens RP, Siriranthsinghji DJS, Kuehn MR, Dunnett SB. Behavioral and neurochemical evaluation of a transgenic mouse model of Lesch-Nyhan syndrome. J Neurol Sci 1988;86:203–213.

Gibbs RA, Caskey CT. Identification and localization of mutations at the Lesch–Nyhan locus by ribonuclease A cleavage. Science 1987;236:303–305.

Gibbs RA, Nguyen P-N, Edwards A, Civitello AB, Caskey CT. Multiplex DNA deletion detection and exon sequencing of the hypoxanthine phosphoribosyltransferase gene in Lesch-Nyhan families. Genomics 1990;7:235–244.

Gibbs RA, Nguyen P-N, McBride LJ, Koepf SM, Caskey CT. Identification of mutations leading to the Lesch-Nyhan syndrome by automated direct DNA sequencing of in vitro amplified cDNA. Proc Natl Acad Sci USA 1989;86:1919–1923.

Haldane JBS. The rate of spontaneous mutation of a human gene. J Genet 1935;31:317–326.

Hooper M, Hardy K, Handyside A, Hunter S, Monk M. HPRT-deficient (Lesch-Nyhan) mouse embryos derived from germline colonization by cultured cells. Nature 1987;326:292–295.

Kelley WN, Wyngaarden JB. Clinical syndromes associated with hypoxanthine-guanine phosphoribosyltransferase deficiency. In: Stanbury JB, Wyngaarden JB, Fredrickson DS, Goldstein JL, Brown MS, eds. The metabolic basis of inherited disease, 5th ed. New York: McGraw-Hill, 1983:1115–1143.

King A, Melton DW. Characterisation of cDNA clones for hypoxanthine-guanine phosphoribosyltransferase from the human malarial parasite, *Plasmodium falciparum:* comparisons to the mammalian gene and protein. Nucl Acids Res 1987;15:10469–10481.

Kuehn MR, Bradley A, Robertson EJ, Evans MJ. A potential animal model for Lesch-Nyhan syndrome through introduction of HPRT mutations into mice. Nature 1987;326:295–298.

Lesch M, Nyhan WL. A familial disorder of uric acid metabolism and central nervous system function. Am J Med 1964;36:561–570.

MacGregor GR, Nelson DL, Chang SMW, Caskey CT. Gene replacement therapy: the example of Lesch-Nyhan syndrome. In: Rowland LP, Wood DS, Schon EA, DiMauro S, eds. Molecular genetics in diseases of brain, nerve and muscle. New York: Oxford University Press, 1989:417–425.

Miller AD, Eckner RJ, Jolly DJ, Friedmann T, Verma IM. Expression of a retrovirus encoding human HPRT in mice. Science 1984;225:630–632.

Nyhan WL, Pesek J, Sweetman L, Carpenter DG, Carter CH. Genetics of an X-linked disorder of uric acid metabolism and cerebral function. Pediatr Res 1967;1:5–13.

Rossiter BJF, Edwards A, Caskey CT. HPRT mutation and the Lesch-Nyhan syndrome. In: Brosius J, Fremeau RT, eds. Molecular genetic approaches to neuropsychiatric disease. San Diego: Academic Press, Inc., 1991:97–124.

Seegmiller JE, Rosenbloom FM, Kelley WN. Enzyme defect associated with a sex-linked human neurological disorder and excessive purine synthesis. Science 1967;155:1682–1684.

Stout JT, Caskey CT. Hypoxanthine phosphoribosyltransferase deficiency: the Lesch-Nyhan syndrome and gouty arthritis. In: Scriver CR, Beaudet AL, Sly WS, Valle DV, eds. The metabolic basis of inherited disease. 6th ed. New York: McGraw-Hill, 1989;1007–1028.

Wilson JM, Stout JT, Palella TD, Davidson BL, Kelley WN, Caskey CT. A molecular survey of hypoxanthine-guanine phosphoribosyltransferase deficiency in man. J Clin Invest 1986;77:188–195.

Yang TP, Patel PI, Chinault AC, et al. Molecular evidence for new mutation at the *hprt* locus in Lesch-Nyhan patients. Nature 1984;310:412–414.

CHAPTER 6

Myotonic Dystrophy

Allen D. Roses
Margaret A. Pericak-Vance

HISTORIC AND CLINICAL DESCRIPTION

Myotonic dystrophy (DM) affects multiple organ systems. The name, myotonic dystrophy, results from the myopathy, dystrophy, and myotonia of skeletal muscle. It is the most prevalent human muscular dystrophy, affecting adults and children with an estimated occurrence of 1/10,000. DM is inherited as an autosomal dominant trait. The disease is variably expressed, and individuals may present with symptoms and signs involving many different organ systems (Table 6.1). Penetrance varies with age, and the disease may affect different tissues at different periods of life. A congenital form of DM may occur when the affected child is the product of a DM carrier mother.

The "textbook," or most easily recognized, clinical presentation of DM usually involves a young adult presenting with weakness of hands or mild foot drop who is discovered on examination to have asymptomatic myotonia, or the sustained contraction and depolarization of skeletal muscle in response to a percussive or electrical stimulus. The distribution of myotonia is usually in the hands and the tongue, while weakness involves predominantly distal extremities. A typical facies or appearance may then be recognized, with loss of temporal muscle and slight weakness of lips and mouth with a "hatchet-like" shape (Figure 6.1). More detailed clinical examination frequently identifies involvement of other systems. Electrocardiographic abnormalities are frequent. Subcapsular, punctate iridescent cataracts may be present in middle-aged patients and may develop into mature cataracts.

With the availability of genetic markers that are closely linked to the

147

Table 6.1: Systemic Involvement in Myotonic Dystrophy

Organ or System	Clinical	Diagnostic Signs
Muscle	Myotonia, weakness, dystrophy.	*EMG:* decreased resting membrane potential; repetitive depolarization ("dive bomber" sound). *Pathology:* sarcoplasmic masses, ringed fibers, internal nuclei, frequent; nuclei often in chains; large variation in fiber size.
Cardiac	Bradycardia common; complete heart block frequent; prolonged PR interval.	First degree heart block, bradycardia on ECG; abnormal vectorcardiogram; SA node, right and left bundle branch dysfunction; increased His–Purkinje conduction (His bundle studies) with progressive conduction system abnormalities.
Lens	Posterior subcapsular, iridescent, or scintillating cataracts.	Dust-like cataracts may be visible only on slit lamp exam.
Eye	Decreased vision (independent of cataracts and diabetic retinopathy); diplopia.	Pigmentary disorders of macula keratosis sicca; decreased intraocular pressure; frequent ptosis and ultraocular muscle weakness.
CNS	Mental retardation (especially congenital DM); hypersomnia.	Possible neuronal heterotopias. Suspicious, reticent personality characteristics.
Endocrine	Abnormal carbohydrate metabolism; testicular (and ovarian) atrophy.	Abnormal glucose tolerance with elevated insulin levels. Gonadal fibrosis (pathology), decreased 17-keto steroids (occasional). Decreased metabolic rate, normal thyroid hormone levels.
Integument	Frontal balding.	Calcifying epitheliomas.
Gastrointestinal	Dysphagia, abdominal pain.	Disordered esophageal and gastric peristalsis; dilation of bowel.

Table 6.1 (continued)

Skeletal	Cranial and facial abnormalities; malocclusion of dentition.	Cranial bony abnormalities, hyperostosis of skull (localized or diffuse), small sella turcica, large sinuses, micrognathia.
Respiratory	Hypoventilation; postanesthesia respiratory failure.	Diaphragmatic and intercostal muscle weakness.
Smooth Muscle	Dilation of hollow visious organs and ureters; abnormal bowel motility.	Thinned or interrupted smooth muscle.

gene locus, large family studies have been performed during the past decade. It is now quite apparent that at any given point in time, fewer than a third of gene carriers will be symptomatic. In addition, another third may be undiagnosable using clinical criteria. The latter group includes not only younger presymptomatic individuals but also older obligate heterozygotes, particularly males, who have an affected parent or siblings as well as affected children. These asymptomatic obligate heterozygotes may have no clinical signs of the disease well into their 50s or 60s; in rare cases they are asymptomatic well into their 80s or 90s.

DM is named for its effects on skeletal muscle. Myotonia can be found in the small muscles of the hand and in the tongue. Repetitive discharges with gradual and uneven decay of amplitude are found using electromyography (Figure 6.2). Myotonia is not often clinically symptomatic as it may be in congenital myotonia, a distinct neuromuscular disorder not charcterized by dystrophy. In DM, muscles affected by myotonia are in the same distribution as the muscles that undergo dystrophic changes (Figure 6.3). The progession of myotonia to severe dystrophy may be quite prolonged, spanning several decades. In general, the younger the presentation, the more rapid the progression. Only a small fraction of affected patients (< 10%) progress to wheelchairs. Many more may benefit from lightweight polyurethane braces worn in shoes to allow ambulation in the presence of foot drop.

The most serious complications are those involving the heart. Cardiac conduction abnormalities are common and may be progressive, particularly in younger patients. Such patients can be identified by periodic EKG monitoring and are usually without obvious cardiac complaints. Sudden death in teenagers involved in athletics is not unusual in DM families. In family studies cardiac signs may be the initial clinical manifestation of the disease. Bradycardia and first-degree heart block are common. When following *DM*

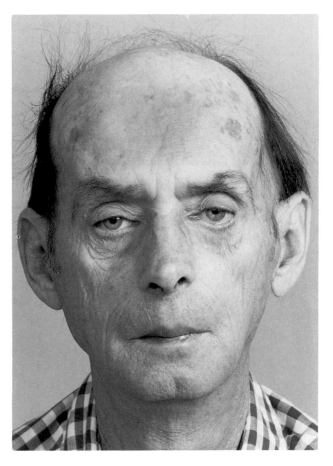

Figure 6.1: A 41-year-old man with DM. Muscle-wasting of temporalis muscles with narrow small chin produce a "hatchet-like" facies. Baldness and ptosis (note droopy eyelids with pupils partially covered and sclerae visible) contribute to characteristic appearance.

gene carriers periodically with electrocardiograms, progression to first-degree heart block (PR interval ≥ 0.20 sec) should lead to other diagnostic tests. In particular, His-bundle ventricular electrophysiologic studies have demonstrated progression of the conduction defects over several years. Complete heart block is not tolerated well in DM because of baseline bradycardia, and can lead to death. Therefore, in the presence of progressive

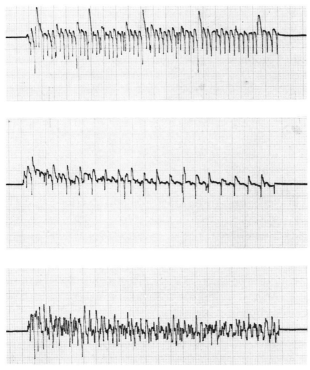

Figure 6.2: Myotonic discharges recorded with an EMG needle in the biceps muscle of a patient with DM. Note spontaneous discharges of variable amplitude and characteristic myotonic discharge in bottom panel.

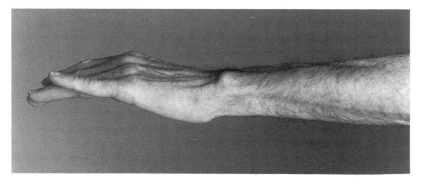

Figure 6.3: Characteristic semiflexed fingers in extended hand of a DM patient. Weak extension and characteristic flexor myotonia is prominent in this patient.

conduction abnormalities and a prolonged or lengthening His-ventricular conduction time, elective insertion of a demand pacemaker may prove to be life-saving.

Other prominent clinical manifestations include central nervous system problems. DM patients may have a particular psychological profile. They may be indifferent, unconcerned, reticent, or hostile. There may be mild retardation compared with unaffected siblings, which is especially prominent in patients with congenital DM. Young and middle-aged affected individuals may be hypersomnolent and indolent, sometimes sleeping up to 20 hours per day. Even in DM patients who maintain a higher level of activity, hypersomnolence can be a prominent symptom and interfere with employment. In family pedigree studies, it is not uncommon for unaffected family members to respond to an invitation to appear at a meeting place for evaluation while affected individuals do not bother or care to attend. Home visits are frequently necessary and, even then, cooperation can be marginal.

Congenital DM (CDM) presents as a unique syndrome that occurs in fewer than 20% of the births involving an affected infant from a heterozygous mother. The mother may be asymptomatic at the time, and, as frequently happens, a branch of a DM family may be ascertained by the birth of a CDM child.

CDM presents at birth with hypotonia (floppy infant), neonatal respiratory difficulties related to poor motor function, absent or weak suck, causing feeding difficulties, talipes, and a characteristic tented or pursed appearance of the mouth due to bilateral facial weakness. Retrospectively, there is usually a history of decreased fetal movement and polyhydramnios. Motor development is usually delayed and mental retardation is common as the child develops. Myotonia is absent during the neonatal period and infancy, but develops over the early childhood years. Muscle tissue that has been examined pathologically is usually described as immature, with abnormal architectural arrangement of myofibrils. CDM is only observed when the affected infant (male or female) is born to an affected mother. In many cases the affected mother is diagnosed after the birth of a CDM infant. Some mothers have no symptoms or signs. While less than 20% of affected children of affected mothers exhibit CDM, if a mother has had one child with CDM then the chances are much more likely (up to 80%) that another affected child would also exhibit this severe form of DM.

Although infantile or childhood onset of signs of DM can be observed in families where the father is affected, CDM is not observed. There have been several speculations over the years about the cause of this phenomenon, but it is still unknown. Some possible explanations include circulating factors from the mother affecting the pregnancy, abnormal opposing fetal and maternal membranes causing transport problems to the infant, a result of maternal mitochondrial inheritance, or DNA imprinting. The exclusive maternal inheritance of CDM may yet provide a clue to the inherited mutation(s). The recurrence of CDM in successive pregnancies in certain carrier mothers should also be kept in mind as potential mechanisms are tested.

MOLECULAR GENETICS

Linkage

DM was the first human autosomal dominant disease for which genetic linkage was established. In 1954, Mohr published evidence that *DM* was linked to the Lutheran (LU) blood group and the ABH secretor locus (*FUT2*). These data were later confirmed by Renwick et al. Since *FUT2* could be measured in amniotic fluid, it could be used in prenatal testing. Thus, DM was one of the first examples of how linkage studies could be applied in clinical situations.

Subsequently, the *DM-FUT2-LU* linkage group was shown to be linked to the third component of complement (C3). Following its cloning and use in linkage studies, C3 was localized to chromosome 19 using somatic cell hybrid techniques. The linkage relationship between *DM* and C3 thus allowed *DM* to also be localized to chromosome 19.

Subsequent linkage studies by a number of groups using RFLPs from a variety of loci provided fine localization of *DM* to the 19q13.2–13.3 region of the chromosome. These markers included the apolipoprotein C2 gene (*APOC2*), which was shown to be tightly linked to *DM* at a distance of 2–4 cM, and the gene for creatine kinase muscle form (*CKM*). Both these markers have proven invaluable in DM prenatal and carrier detection studies since they are tightly linked to *DM* and highly polymorphic. Both *APOC2* and *CKM* map proximal to *DM*. Analysis of crossover and disequilibrium data provided additional mapping information in this region, and established *ERCC1* and *D19S63* as two of the closest proximal markers to *DM*. A long-range restriction map of the region resulted in the order *CKM-ERCC1-D19S63*-qter. A cosmid walk by Korneluk, deJong, and coworkers enabled additional mapping of this region, as well as the generation of the marker *D19S63*, which was then the closest proximal marker to *DM*.

Distal to *DM*, the markers *D19S50* and *D19S22* were shown to flank the *DM* locus at distances of approximately 9 and 13 cM. Subsequently, a random DNA probe D19S51, developed from a radiation hybrid cell line designated 20XP3542-1-4, was also shown to be linked tightly to *DM* and distal to *CKM*. D19S51 was mapped using a hybrid cell line that contains the region of chromosome 19 from a breakpoint in *CKM* through qter. D19S51 was located on that hybrid and thus was defined as distal to the *CKM* gene. Shutler et al. reported a recombination event in a French Canadian family that supports the placement of D19S51 as distal to *DM*. This same individual crossed over with the marker *D19S112*. Physical and genetic data showed that *D19S112* was proximal to D19S51 and thus was the closest distal marker. Studies also indicated that the distance between two of the closest flanking *DM* markers, *D19S118* and *D19S112*, which were linked by a cosmid walk, did not exceed 300 kb.

Although the *DM* gene has been localized for approximately a decade, the putative molecular defect has only recently been defined. Recognition of the site of the mutation was possible by the finding of a CTG variable repeat while screening for highly polymorphic microsatellite probes. This triplet CTG repeat is highly variable in normal individuals, with a range of 5–27 copies reported in one of the initial papers. It was noticed that the triplet repeat could not be amplified from some *DM* DNA samples using PCR techniques that worked in nonaffected individuals. Using Southern blot hybridization techniques, it was found that many DM patients had very high amplifications of the repeat. Minimally affected individuals had more than 50 repeats, while more severely affected individuals in the same family contained repeat segments up to several kilobases in length. Using probes derived from the triplet repeat region, Brook et al., and subsequently Fu et al. and Mahadevan et al., isolated a putative human *DM* gene containing the variable CTG triplet in its 3′ untranslated region. Northern blot and sequence analysis showed this gene to be approximately 3.3 kb in baboon tissues and to possess homologies with members of protein kinase families. The gene was apparently highly expressed in cardiac muscle, expressed in skeletal muscle, and minimally expressed in brain, although the first clone reported was derived from a brain library.

The involvement of a highly variable triplet repeat as a potential regulator of the severity of the disease, and presumably the expression of a mutant product, has important implications in the understanding of variable penetrance and expressivity. Whether the variability is introduced as a meiotic or mitotic event remains to be examined. Whether each tissue in an affected patient contains the same variable insert, and whether the insert size is directly related to tissue expressivity, will soon be known. The role of a huge variable insert in the expression of several genes in the immediate vicinity of the repeat may also play a role in determining the severity of phenotypic expression. Finally, the role of the repeat in the transmission of CDM will need to be examined. Preliminary evidence shows that relatively mildly affected mothers of severely affected DM patients may themselves have very large repeat sequences. The maternal transmission of CDM may have a DNA basis, although the mechanism of expression of the operative gene(s) and products awaits elucidation. Effects on imprinting of DNA, expression of several genes in the region, or metabolic events derived from the mutated or interrupted gene may be involved.

Rapid progress can be expected to be made regarding the mechanism of expression of DM and, with the rapid rate of research information becoming available, remarkable progress in understanding DM will be made before this chapter is published.

MUSCLE PATHOLOGY

In adults with DM, the skeletal muscle pathology generally starts with type 1 fiber atrophy, a finding that has long engendered controversy over whether the muscle changes are due to a neuropathic etiology or to a primary muscle defect In fact, both muscle and nerve are probably affected in DM. Clinical clues often come from the earliest symptoms or signs of disease. There is an abnormal architectural organization of myofibrils in muscle from congenital DM patients. Selective type 1 fiber atrophy, which is characteristic of adult muscle, is not yet present, although it occurs later in the course of the disease. Abnormalities in the peripheral distribution and arrangements of the myofibrils and the myofilaments in the muscle fiber are present in both adult DM muscle and congenital DM. Mussini et al. have shown that one of the earliest changes in DM fibers is the presence of very high electron density at the I band. These phenomena need to be explained in light of the recent molecular advances in DM.

PATHOPHYSIOLOGY

The mutation responsible for DM will ultimately need to address the multisystemic involvement, particularly at the His–Purkinje cell system in the heart, mild nonprogressive mental retardation, characteristic cataracts, etc. Certainly an elusive, ubiquitous, circulating factor can always be invoked. An isoform of a protein that is localized to the appropriate anatomy and provides pathophysiologic mechanism should also be considered. While there is no molecular explanation for the multiple pathophysiologic abnormalities in DM, this situation may change very quickly with the recent progress made in defining chromosome 19.

The electrophysiologic abnormalities of skeletal muscle have suggested the potential involvement of a sarcolemmal membrane ion channel. In congenital myotonia and goat myotonia there is an alteration in chloride flux. This is not present in DM. A decreased resting membrane potential has suggested an alteration in sodium flux. Within muscle, a physiologic abnormality may be located between the membrane and the regulation of calcium.

THERAPEUTIC OPTIONS AND DIRECTIONS

At the present time, treatment of the disease is based on those symptoms and signs that are debilitating or fatal. There are drugs that can lessen symptomatic myotonia, including phenytoin, quinidine, and procainamide. All of these drugs can affect the cardiac system and, since myotonia is seldom more than a minor complaint in DM, treatment is not commonly necessary.

The major therapeutic intervention concerns the anticipatory treatment of potentially lethal cardiac arrythmias. A history of sudden death is very common when attention is focused on large pedigrees. Unfortunately, sudden death frequently occurs in individuals at risk for the disease without the benefit of diagnostic examination. We have approached this lethal complication from a prospective evaluation of all pedigree members under our care. While sudden death may occur for a variety of specific cardiac arrythmias, we have been able to follow the progression of complete bundle branch block in DM patients.

Patients are routinely screened with electrocardiograms. Bradycardia and a prolonged PR interval are relatively common findings. Over time the PR interval can become longer, and second-degree as well as complete bundle branch block can occur. Complete block in an otherwise normal individual may produce an idioventricular rate compatible with life; in DM patients the idioventricular rate is also bradycardiac, frequently fewer than 20 beats per minute, and may produce sudden death. Follow-up ECGs are performed every two to three years and, if a patient extends his PR interval to greater than 0.20 seconds, we will follow-up with an invasive catheterization study of the His-ventricular conduction system. If the H-V conduction time exceeds 60 msec, we will prescribe the placement of a demand cardiac pacemaker. In several cases followed over the past two decades, the demand pacemaker has taken over function when complete heart block has occurred. This serious complication of DM can be anticipated and treated. The rate of change of the electrocardiogram can be quite variable, like every other clinical manifestation of DM. In one teenage girl who was a congenital DM patient and who was studied annually, we found a prolonged PR interval and scheduled her for a His-bundle ventricular study. Between the last clinic visit and the scheduled study she died suddenly in her sleep. Other patients have carried demand pacemakers for more than five years before they have activated.

The dystrophy and weakness associated with DM has no curative treatment. In most cases the weakness develops gradually over many years. Lightweight polyurethane braces for foot drop or weakness of foot dorsiflexor muscles may help ambulation dramatically and be effective for many years. Similarly, physical medicine intervention for individual problems of weakness (also quite variable) can be of considerable help to the patient.

Future directions for treatment of DM will be directed by the knowledge of the gene defect. The disease is expressed over a long time period and this leaves a therapeutic window for treatments designed to control gene expression. Control of abnormal gene expression rather than genetic engineering is likely to become the pharmacology of the future for many neurodegenerative and dystrophic diseases. DM is a superb model since so many systems are involved. Unlike Huntington's disease or Alzheimer's disease, therapeutic testing can involve accessible organs with well-defined anatomy and physiology, such as muscle.

SELECTED REFERENCES

Bartlett RJ, Pericak-Vance MA, Yamaoka LH, et al. A new probe for the diagnosis of myotonic muscular dystrophy. Science 1987;235:1648–1650.

Brook JD, McCurrach ME, Harley HG, et al. Molecular basis of myotonic dystrophy: expansion of a trinucleotide (CTG) repeat at the 3′ end of a transcript encoding a protein kinase family member. Cell 1992;68:799–808.

Brooke MH, Engel WK. The histographic analysis of human muscle biopsies with regard to fibre types. I. Adult and male and female. Neurology 1969; 19:221–233.

Brunner HG, Korneluk RG, Coerwinkel-Driessen M, et al. Myotonic dystrophy is closely linked to the gene for muscle-type creatine kinase (CKMM). Hum Genet 1989;81:308–310.

Caughey JE, Myrianthopoulos NE. Dystrophia myotonica and related disorders. Springfield, Ill: Charles C. Thomas.

Corbo L, Maley JA, Nelson DL, Caskey CT. Direct cloning of human transcripts with HnRNA from hybrid cell lines. Science 1990;249:652–655.

Dodge PR, Gamstorp I, Byers RK, Russell P. Myotonic dystrophy in infancy and childhood. Pediatrics 1966;35:3–19.

Eiberg H, Mohr J, Nielsen LS, Simonsen N. Linkage relationships between the locus for C3 and 50 polymorphic systems: assignment of C3 to the DM-SE-LU linkage group; confirmation of C3-LES linkage; support of LES-DM synteny. Abstract. Sixth World Congress of Human Genetics, Jerusalem, 1981:147.

Farkas E, Tome FMS, Fardeau M, Arsenionunes ML, Dreyfuss P, Dooebler MF. Histochemical and ultrastructural study of muscle biopsy in three cases of dystrophia myotonica in the newborn child. J Neurol Sci 1974;21:273–288.

Fu Y-H, Pizzuti A, Fenwick RG Jr, et al. An unstable triplet repeat in a gene related to myotonic muscular dystrophy. Science 1992;255:1256–1258.

Greenfield JG, Shy GM, Alvord EC, Berg L. An atlas of muscle pathology in neuromuscular disease. Edinburgh: Livingstone, 1957.

Gusella JF, Tanzi RE, Anderson MA, et al. DNA markers for nervous system diseases. Science 1984:225;1320–1326.

Harley HG, Brook JD, Floyd J, et al. Detection of linkage disequilibrium between the myotonic dystrophy locus and a new polymorphic DNA marker. Am J Hum Genet 1991;49:68–75.

Harper PS. Congenital myotonic dystrophy in Britain. 1. Clinical aspects. Arch Dis Child 1978;50:505–513.

Harper PS. Congenital myotonic dystrophy in Britain. 2. Genetic aspects. Arch Dis Child 1978;50:514–521.

Harper PS. Myotonic dystrophy. 2nd ed. Philadelphia: Saunders Pub. Co., 1989.

Johnson K, Shelbourne P, Davies J, et al. Recombination events that locate myotonic dystrophy distal to APOC2 on 19q. Genomics 1989;5:746–751.

Korneluk RG, MacKenzie AE, Nakamura Y, Dube I, Jacob P, Hunter AGW. A reordering of human chromosome 19 long-arm DNA markers and identification of markers flanking the myotonic dystrophy locus. Genomics 1989;5:596–604.

Kunkel LM, Monaco AP, Middlesworth W, et al. Analysis of deletions in the DNA of patients with Becker and Duchenne dystrophy. Nature 1986:322;73–77.

Lipicky RJ. Studies in human myotonic dystrophy. In: Rowland LP, ed. Pathogenesis of human muscular dystrophies. Amsterdam: Excerpta Medica, 1982:729–738.

Lipicky RJ, Bryant SH. Sodium, potassium and chloride fluxes in intercostal muscle from normal goats and goats with hereditary myotonia. J Gen Physiol 1966;50:89–111.

Liu P, Legerski R. Siciliano MJ. Isolation of human transcribed sequences from human-rodent somatic cell hybrids. Science 1989;246:813–815.

Mahadevan M, Tsifidis C, Sabourin L, et al. Myotonic dystrophy mutation: an unstable CTG repeat in the 3′ untranslated region of the gene. Science 1992;255:1253–1255.

Meredith AL, Huson SM, Lunt PW, et al. Application of a closely-linked polymorphism of restriction fragment length to counselling and prenatal testing in families with myotonic dystrophy. Br Med J 1986;293:1353–1357.

Mohr J. A study of linkage in man. Copenhagen: Munksgaard, 1954.

Moorman JR, Coleman RE, Packer DL, et al. Cardiac involvement in myotonic muscular dystrophy. Medicine 1985;64:371–387.

Mussini I, Di Mauro S, Angelini C. Early ultrastructural and biochemical changes in muscle in dystrophia myotonica. J Neurol Sci 1970;10:585–604.

O'Connell P, Viskochil D, Buchberg AM, et al. The human homolog of murine Evi-2 lies between two von Recklinghausen neurofibromatosis translocations. Genomics 1990;7:547–554.

Pericak-Vance MA, Yamaoka LH, Assinder RIF, et al. Tight linkage of apolipoprotein C2 (ApoC2) to myotonic dystrophy (DM) on chromosome 19. Neurology 1986;36:1418–1423.

Prystowsky EN, Pritchett ELC, Roses AD, Gallagher JJ. The natural history of conduction system disease in myotonic muscular dystrophy as determined by serial electrophysiologic studies. Circulation 1979;60:1360–1364.

Renwick JH, Bundey SE, Ferguson-Smith MA, Izatt MM. Confirmation of linkage of the loci for myotonic dystrophy and ABH secretion. J Med Genet 1971;8:407–416.

Ropers HH, Pericak-Vance MA. Report of the committee on the genetic constitution of chromosome 19. Cytogen Cell Genet 1991;58:751–784.

Roses AD, Harper P, Bossen E. Myotonic muscular dystrophy (dystrophia myotonica, myotonia atrophy). In: Vinken PJ, Bruyn CW, eds. Handbook of clinical neurology. New York: John Wiley and Sons, 1979:485–532.

Sahgal V, Bernes S, Sahgal S, Lischwey C, Subramani V. Skeletal muscle in preterm infants with congenital myotonic dystrophy. J Neurol Sci 1983;59:47–55.

Sarnat HB, Silbert SW. Maturational arrest of fetal muscle in neonatal myotonic dystrophy. Arch Neurol 1976;33:466–474.

Schrott HG, Karp L, Omenn GS. Prenatal prediction in myotonic dystrophy: guidelines for genetic counseling. Clin Genet 1973;4:38–45.

Shaw DJ, Meredith AL, Sarfarazi M, et al. The apolipoprotein CII gene: subchromosomal localization and linkage to the myotonic dystrophy locus. Hum Genet 1985;70:271–273.

Shutler G, MacKenzie AE, Brunner H, et al. Physical and genetic mapping of a novel chromosome 19 ERCC1 marker showing close linkage with myotonic dystrophy. Genomics 1991, in press.

Smeets H, Backinski L, Coerwinkel M, et al. A long-range restriction map of the human chromosome 19q13 region: close physical linkage between CKMM and the ERCC1 and ERCC2 genes. Am J Hum Genet 1990;46:492–501.

Smeets HJM, Hermens R, Brunner HG, Ropers H-H, Wieringa B. Identification of variable simple sequence motifs in 19q13.2-qter:markers for the myotonic dystrophy locus. Genomics 1991;9:257–263.

Speer MC, Pericak-Vance MA, Yamaoka L, et al. Presymptomatic and prenatal diagnosis in myotonic dystrophy by genetic linkage studies. Neurology 1990; 40:671–676.

Thomson AMP. Dystrophia cordis myotonica studied by serial histology of the pacemaker and conducting system. J Pathol Bacteriol 1968;96:285–295.

Wallace MR, Marchuk DA, Andersen LB, et al. Type 1 neurofibromatosis gene: identification of a large transcript disrupted in three NF1 patients. Science 1990;249:181–186.

Whitehead AS, Solomon E, Chambers S, Bodner WF, Povey S, Fey G. Assignment of the structural gene for the third component of human complement to chromosome 19. Proc Natl Acad Sci USA 1982;79:5021–5025.

Yamaoka LH, Pericak-Vance MA, Speer MC, et al. Tight linkage of creatine kinase (CKMM) to myotonic dystrophy on chromosome 19. Neurology 1990; 40:222–226.

CHAPTER 7

Neurofibromatosis

Margaret R. Wallace
Francis S. Collins

Neurofibromatosis (NF) is a term encompassing at least two distinct human genetic diseases in which the common feature is growth of tumors arising from nerves. Due to great variability in the symptoms, there has historically been confusion about classification of these syndromes, leading to delineation of numerous types of NF. The most recent consensus, however, with the help of some stringent criteria recently developed, has categorized neurofibromatosis into two predominant types, NF1 and NF2. Whether or not the other NF forms described are variants of these remains to be settled by molecular genetic studies. NF1 (von Recklinghausen or peripheral neurofibromatosis) and NF2 (bilateral acoustic or central neurofibromatosis) are now known to be completely separate diseases by virtue of genetic linkage of NF1 to chromosome 17q and NF2 to chromosome 22q. Part of the *NF1* gene was recently cloned, and shows significant homology to GAP proteins. Thus, its function may be involved in regulation of GTP-associated proteins, such as *ras,* which are known to be involved in cell growth. There has been no evidence of genetic heterogeneity (multiple loci) for either form of neurofibromatosis. For the purpose of this chapter, only classical NF1 and NF2 will be discussed, and each will be described separately.

GENETICS OF NF1

The first clear description of NF1 and the identification of neurofibromas arising from neural tissue was published in 1882 by Frederick von Recklinghausen in Germany, but presumably the disease has been around much longer. Since then, many patients and families have been described, but relatively little is known about the biology of the disease. It was originally thought that Joseph Merrick, the "Elephant Man," was afflicted with NF1 and thus it became known as "The Elephant Man Disease." However, it is now believed that Merrick instead had the Proteus Syndrome, leaving NF1 with an erroneous association.

NF1's mode of inheritance, which was initially unclear due to variable expressivity (even within families) and a high mutation rate, was determined to be autosomal dominant. The mutation rate has been estimated at 10^{-4} mutations per allele per generation, based on new mutations being responsible for 30–50% of NF1 cases, although there has been some speculation that the mutation rate is actually somewhat lower. One past hypothesis explaining the high mutation rate was that NF1 was heterogeneous (i.e., had multiple loci). However, all linkage studies to date indicate that there is only one locus, in a limited region of 17q11.2, refuting that argument. The fact that the recently identified *NF1* gene is quite large (encoding a transcript nearly the size of the dystrophin mRNA) makes it likely that this locus is a large target for mutational events, as seen in Duchenne muscular dystrophy and retinoblastoma.

Despite the complicating factors of variable expressivity and new mutation, careful studies showed that NF1 is very highly penetrant, with no documented cases of nonpenetrance in adults. While there is debate about whether there is a paternal age effect, studies of new mutations indicate that the majority of mutations are of paternal, not maternal, origin. NF1 is also equally predominant in all ethnic groups and is one of the most common genetic diseases, occurring in approximately 1/3,000 births. Whether it is a true dominant disease (i.e., the same phenotype in homozygotes and heterozygotes) is unknown, and is a question that remains to be answered by molecular genetic analysis of the offspring of rare NF1 × NF1 matings.

The hallmarks of the disease (described in the next section) become more evident with age, such that nearly all postpubertal individuals display symptoms that meet diagnostic criteria. Thus, if an individual with no previous NF family history is diagnosed with NF1, one can usually determine if his or her disease is due to a new mutation by very careful examination of the parents. Accurate diagnosis has been essential for genetic linkage studies. Unfortunately, the gradual onset of NF1 often makes diagnosis in young children difficult.

CLINICAL ASPECTS OF NF1

Café-au-Lait Spots

The standard features for definitive diagnosis of NF1 are described here and discussed in several reviews, and the accepted diagnostic criteria are shown in Table 7.1. The most common sign (present in more than 95% of patients), and often the first to appear, is the presence of pigmented dermal patches called café-au-lait spots, which are the color of creamed coffee (thus their name). Occasionally present at birth, these spots usually become larger and more numerous with age, and are found primarily on the trunk. Café-au-lait spots often show an increase in the rate of appearance and growth during puberty or pregnancy, perhaps indicating a sensitivity of these lesions to sex hormone influence. These spots can occur in normal individuals, but rarely are there more than one or two. Several kindreds showing only dominantly inherited multiple café-au-lait spots have been described, and café-au-lait spots are also seen in other disorders, so these spots alone are not diagnostic for NF1. But the presence of six or more spots, having a diameter of at least 0.5 cm prior to puberty or at least 1.5 cm in postpubertal patients, is a very strong sign, and is one of the diagnostic criteria for NF1. Freckling in the axillae and groin, uncommon locations for freckles in normal individuals, is also a consistent sign of NF1. Any individual meeting at least two of the criteria listed in Table 7.1 is considered to have NF1.

Neurofibromas

Cutaneous neurofibromas are another common feature of NF1 and give the disease its name. These benign tumors (usually soft, painless nodules) appear to arise from peripheral nerve sheaths and contain nerve axons, fibroblasts, Schwann cells, vascular elements, and mast cells. Although they may be present at any age, these often appear during adolescence and increase in number and size with age. Cutaneous neurofibromas rarely

Table 7.1: Diagnostic Criteria for NF1

Two or more of the following:
Six or more café-au-lait spots:
   1.5 cm or larger in postpubertal individual
   0.5 cm or larger in prepubertal individual

Two or more neurofibromas of any type, or at least one plexiform neurofibroma.

Freckling of the armpits or groin.

Optic glioma.

Two or more Lisch nodules.

Dysplasia of the sphenoid bone, or dysplasia or thinning of long bone cortex.

First-degree relative with NF1.

become malignant, but can impede the functioning of surrounding tissues and are disfiguring. An early study indicated that the tumors are of polyclonal origin, but the possibility still exists that one of the cellular elements is clonal.

The other, less common type of neurofibroma is termed plexiform, and is distinguished from cutaneous neurofibromas by its occurrence in deeper tissues (arising from major nerves) and large size and histologic complexity. Plexiform neurofibromas occur at any stage in life, and especially in childhood may cause bone deformities or otherwise impair function of adjacent tissue. These tumors have a greater propensity for malignant transformation than the cutaneous form, with a significant but not well-quantitated risk of developing into neurofibrosarcomas.

Lisch Nodules

Another standard feature of NF1 is the presence of Lisch nodules, which are hamartomas of the iris. A slit lamp examination is necessary to distinguish these from simple pigmented spots on the iris. The presence of at least two Lisch nodules meets one criterion for NF1 (Table 7.1), and it is estimated that over 90% of adult NF1 patients display this feature. Lisch nodules have no known functional effect on the eye.

Other Features

There are a variety of other, less common features of NF1, which include macrocephaly, bone dysplasias, scoliosis, learning disabilities or mental retardation, seizures, hypertension, constipation, pruritus, headaches, short stature, and malignancy (especially of the nervous system). In general, approximately two-thirds of NF1 patients are able to live relatively normal lives, though suffering from cosmetic disfigurement and requiring careful medical attention. There is no consistent association of severity with any known variable (such as sex, new mutation, or the number of generations through which the gene has been transmitted). In addition, severity shows no familial tendency; severely affected parents have no greater likelihood of having severely affected children than mildly affected parents. Thus, it is very difficult to predict the course of the disease in any individual, especially children. Patients who do not develop neoplasms or suffer tissue compromise from neurofibromas often have normal life spans. However, the overall survival rate of NF1 patients is somewhat lower than normal due to the increased incidence of cancer and life-threatening complications.

Malignancy

The question of increased malignancy in NF1 patients has not been adequately answered. While some studies report an increased overall cancer rate, other studies find fault with ascertainment and question the reported increase in neoplasm. It is generally accepted, however, that NF1 patients have at least a slightly increased risk of neural malignancy, including neurofibrosarcomas, schwannomas, astrocytomas, optic glioma, and meningioma. One careful study indicates that NF1 patients may also have a modest increase in non-neural solid tumors as well. In some studies, cultured skin fibroblasts from NF1 patients demonstrated an increased tendency toward transformation by the Kirsten murine sarcoma virus. However, other studies have failed to show such a tendency with other viral agents or by transfection with oncogenes. Thus, the mechanisms of tumor formation and progression to malignancy are unclear.

Associations

Several studies have proposed that certain other disorders tend to occur more frequently in NF1 patients than in the general population. These include Noonan syndrome, rare childhood leukemias, myotonic dystrophy, and Charcot-Marie-Tooth disease (CMT). Since the myotonic dystrophy gene resides on chromosome 19, the two diseases are not syntenic, and so any association is likely to be related through other mechanisms. Noonan syndrome is a variably expressed dominant disease that shares some features with NF1. Although some have postulated that Noonan syndrome may be linked to NF1 or perhaps is an allele of the same gene, there is at least one report of independent segregation of the two diseases. The reported propensity to leukemia is under debate, and some suggest that if true, it may be related to the increased risk of certain malignancies in NF1 (although the hematopoietic system is not of neural crest origin).

The observed coincidences of NF1 and CMT are interesting in light of the recent genetic linkage of the most frequent form of CMT to chromosome 17p, thus placing it in relatively close proximity to the *NF1* gene. While this could explain cosegregation, the occurrence of these two disorders in a single individual is unlikely to be due to a contiguous gene deletion (since this would cross the centromere). No cytogenetic or molecular evidence of pericentric inversion of chromosome 17 has been reported in these patients, but this bears further study.

Segmental Neurofibromatosis

A number of patients have been identified with segmental rather than constitutional NF1. These individuals have café-au-lait spots and neurofibromas localized to one region of their body (usually left or right upper quadrant and adjoining limb), usually not crossing the midline. Lisch nodules are rarer in these patients than in classical NF1. Parents are uniformly unaffected, and therefore postzygotic somatic mutation in early embryogenesis has been proposed as a mechanism for this phenomenon. Presumably the patient's germ line can carry the mutation if the gonadal tissue developed from the cells that suffered the mutation, and thus affected children would occasionally have the disease constitutionally; this in fact has been observed. But germ-line status is difficult to determine, and thus estimating the risk to offspring of these patients is empirical.

MOLECULAR GENETICS OF NF1

Genetic Linkage

Early attempts to localize the *NF1* gene employed classical genetic linkage studies, which were improved considerably in the 1980s by the following: (1) the advent of DNA markers (RFLPs) to test linkage throughout the genome; (2) the establishment of diagnostic criteria that led to more accurate phenotyping, in turn minimizing the error in linkage studies; and (3) the establishment of collaborations such as the International Consortium for Neurofibromatosis Linkage Analysis to promote exchange of probes and family resources and to pool data. Several organizations, such as the National Neurofibromatosis Foundation, have been involved in organizing and funding these efforts.

Some of the earliest linkage data, presented at a 1985 conference, reported the exclusion of certain regions of the genome and certain candidate genes (such as some oncogenes). This was followed by a meeting in England in February of 1987, at which time a much more significant portion of the genome was excluded. The conclusion of those studies was that *NF1* most likely resided on chromosome 5, 10, 17, or 18, with 17 being the most likely candidate. Shortly thereafter, two groups reported linkage of NF1 to markers on chromosome 17. Laboratories then developed more probes and pooled their linkage data, which was presented at an NF1 conference in November 1987, and is documented in a series of papers in *Genomics* (1987;1:337–383). Several potential NF1 candidate genes, such as the oncogene *erb*-A1, were ruled out by detection of genetic recombination. Genetic heterogeneity was also found to be extremely unlikely.

With this continued cooperation among laboratories, more probes were defined and more detailed linkage data obtained. Another NF1 conference was held in June 1988. The pooled data, along with individual group reports, was published in the *American Journal of Human Genetics* (1989;44:1–72). These efforts resulted in unequivocal assignment of NF1 to proximal 17q and development of a genetic map of the

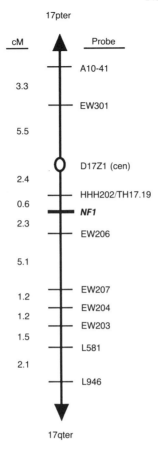

17pter

cM	Probe
	A10-41
3.3	
	EW301
5.5	
	D17Z1 (cen)
2.4	
	HHH202/TH17.19
0.6	
	NF1
2.3	
	EW206
5.1	
	EW207
1.2	
	EW204
1.2	
	EW203
1.5	
	L581
2.1	
	L946

17qter

Figure 7.1: Map showing genetic distances (in cM) between chromosome 17 polymorphic probes linked to NF1. Chromosome 17 is represented by the vertical line, with distances shown on the left representing cM between the probes on the right. The distance between the nearest centromeric probe complex HHH202/TH17.19 and the *NF1* locus is 0.6 cM, and the distance between *NF1* and the nearest distal probe, EW206, is 2.3 cM. All other distances are between adjacent probes, with the order being determined by multipoint linkage analysis.

flanking markers (Figure 7.1). At this point, prenatal and/or presymptomatic diagnosis became possible for families with sufficient pedigree structure and information.

As the linkage efforts proceeded, several groups also began developing physical maps of chromosome 17 for further localization and ultimately for efforts to clone the gene. Two sets of somatic cell hybrids proved to be quite useful in localizing and ordering probes. One set included standard rodent fusion hybrids made from patients with chromosome 17 translocations not associated with NF1. These hybrids provided physical markers at the centromere as well as at several points on 17q, allowing the chromosome to be subdivided (Figure 7.2). The second set of

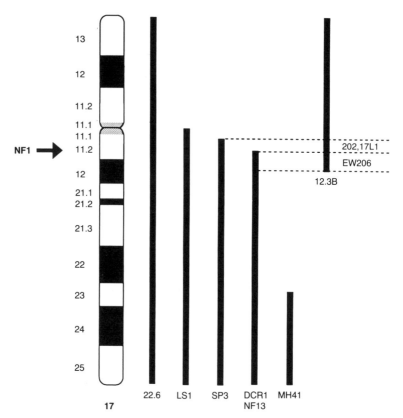

Figure 7.2: Schematic showing human chromosome 17 sequences present in a series of somatic cell hybrids. The solid vertical lines represent the portion of the chromosome present in each hybrid, whose names are shown below the lines. The *NF1* gene location is indicated in 17q11.2, at the end point of the two hybrids (DCR1, NF13) whose breakpoints are cytogenetically indistinguishable. The hybrids allow the chromosome to be subdivided physically. The region just centromeric to *NF1*, between SP3 and DCR1/NF13, is delineated by dashed lines, and two important probes, HHH202 and 17L1, reside in this region. Similarly, the closest distal genetic probe, EW206, maps to the adjacent telomeric region (shown by dashed lines), between DCR1/NF13 and 12.3B. All of these hybrids have mouse backgrounds.

hybrids was derived from microcell fusion of a chromosome 17, tagged with a neomycin resistance gene, in which each hybrid contained a small fraction of the chromosome, often from proximal 17q as well as the region containing the thymidine kinase gene (distal 17q, under selection in the fusion). A panel of these "reduced" hybrids proved useful in ordering closely linked probes proximal to *NF1*. Most reassuring was that the physical mapping and genetic linkage data closely agreed. The combined efforts of these two approaches placed *NF1* in the region of 17q11.2, flanked most closely on the proximal side by a nonpolymorphic clone called 17L1 as well as the genetic marker *HHH202* (1 centiMorgan, cM); distally the closest marker was *EW206* (3 cM) (Figure 7.1).

Translocations

Though the traditional approach to studying autosomal dominant disease does not include karyotyping, limited efforts have been underway to karyotype NF1 patients in search of large chromosomal rearrangements such as deletions or translocations. Such rearrangements can provide crucial chromosomal landmarks that are extremely helpful in physical mapping; for example, cytogenetic abnormalities first allowed mapping of the retinoblastoma and Duchenne muscular dystrophy loci and were very important in the actual gene cloning and analysis.

Two NF1 patients have been identified who possess constitutional reciprocal autosomal translocations involving 17q11.2. One patient carried t(1;17) with clear coinheritance of NF1 and the translocation in her family, and the other patient carried t(17;22). The breakpoints fell in the *NF1* region previously identified by physical and genetic mapping, supporting the hypothesis that these breaks occurred within or near the *NF1* gene and were responsible for the NF1 phenotype in these patients.

The first step in further localizing these was the construction of somatic cell hybrids from these patients, in which the translocation product carrying 17q11.2-qter was the only chromosome 17 material present (Figure 7.2). The two translocation hybrids [DCR-1 for t(1;17) and NF13 for t(17;22)], whose 17q11.2 breakpoints are cytogenetically indistinguishable, thus provided two crucial physical landmarks that, as was later determined, marked the location of the *NF1* gene. This allowed classification of probes as proximal or distal to *NF1*, and efforts shifted toward cloning these breakpoints. An additional chromosome 17 cytogenetic rearrangement was identified in an NF1 patient; this individual has a deleted chromosome 17 (breakpoints in centromere and 17q12) in which the deleted portion exists as a mini-chromosome.

The initial approach to identification of the two translocation breakpoints employed pulsed field gel electrophoresis, which allows examination of several hundred kilobases (kb) of DNA at one time. Briefly, DNAs from the two patient cell lines and normal controls were digested with rare cutting enzymes and electrophoresed on pulsed field gels. These gels were blotted and hybridized with probes in the *NF1* region. The detection of both *NF1* translocation breakpoints by pulsed field gels was achieved by two groups in 1989. One group developed 17L1, which is a linking clone from a *Not* I linking library constructed from chromosome 17 material. This probe (proximal to the breakpoints and encompassing 7 kb spanning the *Not* I site; Figure 7.3) clearly detected the t(1;17) break and less clearly showed evidence of the t(17;22) break. Another group detected these breaks with a cosmid called 1F10, which lies just distal to *NF1* (Figure 7.3). These reports provided a detailed physical map of the *NF1* region and placed the t(1;17) break closer to the centromere. More recent detailed mapping indicates that the breakpoints are only 60 kb apart.

Cloning of the *NF1* Gene

The 60-kb region of the translocation breakpoints was the object of the intensive search for the *NF1* gene. This entire region was cloned in overlapping cosmids and these cosmids, or subfragments, were used to isolate candidate cDNAs. In more novel technology, chromosome jumping and yeast artificial chromosomes (YACs) were also employed to clone and analyze this region.

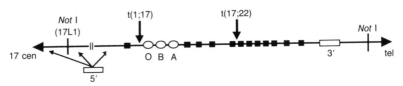

Figure 7.3: Schematic of the genomic region containing the *NF1* gene. The *Not* I sites that define the 350-kb fragment are indicated. The centromeric *Not* I site marks the location of the 7-kb linking clone 17L1. The characterized *NF1* gene exons are indicated by solid boxes, with their size and spacing not drawn to scale. The open box labeled 3′ indicates that the extreme 3′ end of the gene remains to be cloned and thus the number and distribution of exons is unknown. Likewise, the exons at the 5′ end of the gene, represented by the open box labeled 5′, remain to be characterized. The arrows from this box indicate that the exact position of the 5′ end remains to be defined. The locations of the translocation breakpoints, which are 60 kb apart, are indicated. The ovals next to t(1;17) represent the three embedded genes: O represents *OMGP*, B represents *EVI2B*, and A represents *EVI2A*. Again, the size and spacing is not to scale; however, the order is correct, with all three genes being transcribed toward the centromere (i.e., on the *NF1* antisense strand).

The first candidate cDNA, *EVI2A,* was the homolog of a mouse gene involved in virally induced murine leukemia. This gene was identified as an *NF1* candidate by virtue of its location between the breakpoints (Figure 7.3). While the function of this 1.9-kb transcript has not been discovered, extensive studies failed to reveal any mutations in NF1 patients, thereby discounting *EVI2A* as the *NF1* gene.

The second candidate gene, *EVI2B,* was cloned by cosmid and jump clone screening of cDNA libraries. This gene, also completely between the breakpoints, encodes a 2.1-kb mRNA and lies only 4 kb centromeric to *EVI2A* (Figure 7.3). Like *EVI2A,* the function of *EVI2B* is unknown and no mutations have been identified in NF1 patients. Both of these genes are transcribed toward the centromere and have a similar structure of one small 5′ untranslated exon and one large 3′ exon.

Another gene, *OMGP* (oligodendrocyte myelin glycoprotein), lies between the breakpoints and is very close to t(1;17). It encodes a 433–amino acid protein expressed in oligodendrocytes and Schwann cells, making it a strong *NF1* candidate gene. However, detailed patient analyses have also failed to reveal mutations, making *OMGP* the third candidate gene to fail the test.

The techniques described above finally yielded the true *NF1* gene in July 1990. Its transcript is very large (> 11 kb by Northern blot) and the gene spans both translocation breakpoints (Figure 7.3). It is transcribed toward the telomere, with a significant portion of the gene lying centromeric to the breakpoints. It was unequivocally proven to be the *NF1* gene by virtue of its interruption by the translocations as well as identification of mutations in NF1 patients, including in one instance a new-mutation patient whose 0.5-kb insertion is not found in either parent.

The 5′ end of the gene remains to be cloned and sequenced, but the deduced amino acid sequence of the 5′-most cloned portion shows significant homology to GAP (GTPase activating proteins) in the protein data bases, particularly with the yeast GAP genes IRA1 and IRA2. This homology is the first hint of a function for the *NF1* protein—it may act as a GAP protein. GAP proteins, including those in yeast, stimulate the hydrolysis of GTP bound to the *ras* oncogene protein. The *ras* product is a cell growth promoter, especially when it contains an oncogenic mutation and is in the active form when GTP is bound. Thus, GAP proteins inhibit the *ras* protein's function by causing conversion of the GTP to GDP, potentially acting as tumor suppressors. This fits very well with the postulated role of the *NF1* gene, which is already hypothesized to encode a tumor suppressor. Future experiments will determine whether the *NF1* protein has GAP activity (by assays or complementation of mutations in other GAP proteins) or interacts with *ras.*

Analysis of the remaining 5′ sequence of the *NF1* gene may reveal other potential functions as well. The genomic extent of this gene has not been fully characterized, but could span hundreds of kilobases, as the 4 kb of 3′ exons alone covers approximately 100 kb in the genome. It also seems likely that, in light of the high mutation rate, every unrelated NF1 patient will have a unique mutation.

Possible Mechanisms in NF1

Since NF1 is characterized by abnormal growth properties of some cell types, particularly those of neural crest origin, the two most-favored hypotheses of its mechanism prior to gene cloning were oncogene activation or tumor-suppressor inactivation. Since the mutations so far identified clearly disrupt the *NF1* gene, it now appears that NF1 results from reduced or absent expression of this large gene. This places *NF1* in the category of tumor suppressor, along with genes such as *p53* and retinoblastoma (*Rb*). Thus, like *Rb*, *NF1* may fit Knudson's two-hit theory, which would postulate that the inherited *NF1* mutation is the first hit in a growth-suppressing gene, and subsequent random somatic mutations in the remaining normal allele eliminate function. A cell harboring such a "second hit" would then have a proliferative advantage, perhaps resulting in a neurofibroma (if the parent cell is a Schwann cell) or a Lisch nodule (if an iris cell) or a café-au-lait spot (if a melanocyte). It is plausible by analogy to other better-characterized systems that mutations at other cancer-pathway loci (such as *p53*) may then lead to malignancy. This may explain the apparently elevated rates of certain types of cancer in NF1 patients.

In studies of other tumor-suppressor genes, such as *Rb* and Wilms tumor, loss-of-heterozygosity (LOH) analyses (which involve typing patient and tumor DNA for RFLPs) have revealed that the tumors have somatically lost the normal allele, in support of the Knudson hypothesis. In NF1, a number of groups have performed LOH studies on malignant tumors such as neurofibrosarcomas (neurofibromas being of little use due to their mixture of cell types) using RFLPs across chromosome 17. Many of these tumors have shown losses for the entire chromosome, with a few having only 17p deleted. No cases of loss only in the immediate *NF1* region have been reported, however. The occasional loss of 17p may reflect events occurring at the *p53* locus, as *p53* mutations have been found by DNA sequencing of several neurofibrosarcomas from NF1 patients. The lack of LOH specific for the *NF1* region is logical, as the process of tumor development is commonly an independent mutation in tumor-suppressor genes. This would not change the zygosity of nearby RFLP markers; given the high germ-line mutation rate in NF1, it is plausible that the somatic mutation rate is also high. Alternatively, the Knudson hypothesis may turn out not to apply to NF1. Now that the gene is cloned, a high priority will be to look for additional somatic mutations in NF1 tumors.

The identification and cloning of the *NF1* gene, in addition to being highly significant in itself, revealed that the other three genes in the region (*EVI2A, EVI2B, OMGP*) lie in one intron on the opposite (antisense) strand. There has only been one other example of this embedded gene phenomenon in humans: a small gene of unknown function was discovered within an intron of the factor VIII gene. The presence of these embedded genes may be coincidence, or may represent a complex gene regulatory system. It is tempting to speculate that the expression of all four genes is linked, and that mutations in the *NF1* gene that alter its expression may perturb the balance of these genes. Such a disturbance in expression of one or more of the internal genes may result in part of the myriad features seen in NF1. Antisense RNA regulation has been well documented in prokaryotes, and might lead to overexpression of the embedded genes if *NF1* expression is lost. Research into the potential role of the embedded genes in NF1 may reveal new characteristics about the packaging and regulation of mammalian genes.

With the cloning of the *NF1* gene, research into the mechanisms of the disease and possible treatments should proceed much faster. Studies of the *NF1* protein will answer questions about its normal function and abnormalities in NF1 patients. Another question to be answered will be why *NF1*, like *Rb*, is expressed ubiquitously but only limited tissues are affected by the disease. Animal models using embryonic stem cells and homologous recombination will likely be pursued, which may lead to a better understanding of NF1.

While prospects for a treatment or cure for NF1 in the near future are uncertain, they are decidedly brighter than prior to the cloning of the gene. The immediate clinical application of having the *NF1* gene in hand will be for diagnosis. The *NF1* genes of individuals who are suspected of having NF1 but do not quite meet NIH diagnostic criteria (especially if prepubertal) can be examined for mutations. Identification of mutations and RFLPs within the gene will also allow the possibility of prenatal diagnosis for those individuals or families. A crucial and as yet unanswered question is the frequency of gross rearrangements in the *NF1* gene, since these will be easier to detect than point mutations.

Recent Discoveries

Since the original submission of this chapter, a number of discoveries in NF1 have been published. The entire cDNA sequence was reported, which showed that the linking clone 17L1 represents the 5′ end of the *NF1* gene, and thus nearly the entire gene is contained within the 350-kb *Not* I fragment shown in Figure 7.3 (Marchuk et al. *Genomics* 1991;11:931–940). This also confirmed the protein homology to GAP.

The GAP-like domain was shown to function as a GAP protein in yeast expression studies (Ballester et al. *Cell* 1991;63:851–859; Xu et al. *Cell* 1991;63:835–841) and to interact with the *ras* protein (DeClue et al. *Molec Cell Biol* 1991;11:3132–3138). Antibodies directed against the protein product of the *NF1* locus, now called neurofibromin, have been successfully raised and detect a protein of apparent molecular weight 250 kD (Gutmann et al. *Proc Natl Acad Sci USA* 1991;88:9658–9662).

Studies of neurofibroma DNA via an X-inactivation pattern gave evidence that neurofibromas are of clonal origin, although the cell type involved was not identified, nor were gene mutations (Skuse et al. *Am J Hum Genet* 1991;49:600–607). Regarding NF1 patient mutations, the insertion described on page 170 was found to be an Alu element inserted into an intron, causing a splicing error (Wallace et al. *Nature* 1991;353:864–866); another large deletion was described (Upadhyaya et al. *J Med Genet* 1990;27:738–741); and a 5-bp deletion plus adjacent transversion was characterized (Stark et al. *Hum Genet* 1991;87:865–867), in addition to the mutations described by the papers in *Cell* by Cawthorn and associates and Viskochil and coworkers (see Selected References). Most mutations in the *NF1* gene appear to be subtle, however, making direct DNA testing in the clinical setting difficult.

INTRODUCTION TO NF2

NF2 is also known as central or bilateral acoustic neurofibromatosis (BANF) due to its hallmark, bilateral Schwann cell neuromas developing from the acoustic nerves. Since NF2 shows autosomal dominant inheritance and may share a few characteristics with NF1, it has been previously hypothesized that NF1 and NF2 might represent alleles of the same gene. However, genetic linkage has confirmed localization of the *NF2* gene to chromosome 22, not 17, indicating that NF2 is an independent disorder.

The first clinical descriptions of NF2, which emphasized the acoustic tumors, can be found in the literature as early as 1822, being first termed "neurofibromatosis" in the early 1900s. NF2 was then classified as "central" in contrast to "peripheral" (NF1) in 1937. Subsequently, investigators have compared many clinical and genetic aspects of NF1 and NF2, and have developed criteria for the differential diagnosis. As has already been accomplished with NF1, researchers are attempting to clone the *NF2* gene using positional cloning (reverse genetics).

GENETICS OF NF2

NF2 was initially determined to be an autosomal dominant disorder by the study of a few very large pedigrees. However, NF2 occurs at a much lower frequency than NF1 (1 per 50,000 persons versus 1 per 3,000 persons). In contrast to NF1, the mutation rate appears to be quite low. As expression of NF2 can be variable, the NIH has also developed a set of diagnostic guidelines for this disease. A patient is diagnosed with NF2 if he/she meets one or both of the following criteria: (1) bilateral eighth-nerve masses seen with appropriate imaging techniques (i.e., MRI or CT scan); and (2) a first-degree relative with NF2, and the patient has either unilateral eighth-nerve mass or any two of (a) neurofibroma, (b) meningioma, (c) glioma, (d) schwannoma, or (e) juvenile posterior subcapsular lenticular opacity. The disease appears to be fully penetrant (a characteristic that has facilitated mode of transmission studies and linkage efforts), although the symptoms may not appear or significantly worsen until later adulthood. NF2 has shown no evidence of genetic heterogeneity, with all informative families showing linkage to chromosome 22.

CLINICAL ASPECTS OF NF2

The classical finding in NF2 is acoustic neuroma, often bilateral, and is seen in 95% of patients by age 30. These Schwann-cell tumors, which arise from the vestibular nerves, usually do not become clinically evident until adulthood. The symptoms include progressive hearing loss, imbalance, tinnitus, and other findings associated with eighth-nerve masses. Although these tumors are benign, they can become large enough to cause death due to intracranial pressure if they are not removed.

The finding of bilateral acoustic neuromas alone is sufficient for diagnosis of NF2. The growth of these tumors is often very slow and variable, though it has been observed that the growth rate may be increased under the influence of sex hormones (e.g., during puberty or pregnancy). It is important to realize that unilateral acoustic neuroma can occur sporadically, with no known relationship to NF2.

Several studies have found that approximately half of NF2 patients also exhibit some café-au-lait spots or neurofibromas, but usually many fewer than seen in typical NF1 patients. Lens opacities are relatively common and occur at an early stage in the disease, but Lisch nodules are not seen. Other central nervous system tumors, such as meningiomas, gliomas, and schwannomas, are also a hallmark of the disorder, especially involving other cranial or cervical nerve roots.

MOLECULAR GENETICS OF NF2

The first studies that provided clues about the genetic location of *NF2* were LOH and cytogenetic studies of meningioma and acoustic neuromas (both sporadic and associated with NF2). These experiments revealed the consistent loss of part or all of chromosome 22 in many of these tumors, as well as in neurofibromas from NF2 patients. Given this data, linkage efforts were concentrated on chromosome 22, and shortly thereafter the disease was localized to 22q. A more detailed multipoint linkage analysis resulted in identification of flanking markers, *D22S1* and *D22S28*, approximately 13 cM apart. Although all families tested in this study show linkage to chromosome 22, other families remain to be included in these studies to rule out genetic heterogeneity definitively. Thus far, there have been no reports of patients with small deletions or translocations that might aid in the cloning of the gene, as was so useful in NF1. Efforts are currently underway to develop a hybrid mapping panel of chromosome 22 as well as use of YAC and PFG technology to close in on the *NF2* gene.

On the basis of the LOH analyses and disease phenotype, the *NF2* gene is almost certainly a tumor-suppressor gene. However, it seems unlikely that the *NF2* gene is as large as the *NF1* gene, since the NF2 mutation rate is much lower. Although the *NF2* gene has not yet been cloned, the flanking markers can be used for preclinical diagnosis in these families. Identification of patients prior to onset of symptoms may allow earlier intervention to minimize the effects of the disease.

SUMMARY

The two major types of neurofibromatosis are NF1 (von Recklinghausen neurofibromatosis) and NF2 (bilateral acoustic neurofibromatosis), with other clinically unusual types potentially being allelic to one of these. The two diseases have some features in common (i.e., autosomal dominant, development of tumors arising from nerve sheaths) but are also strikingly different in terms of mutation rate and can almost always be distinguished clinically. Although they clearly represent mutations at two different loci (*NF1* on chromosome 17 and *NF2* on chromosome 22), it is tempting to speculate that the genes might be related, at least functionally.

The *NF1* gene, recently identified, is very large and has several other small genes embedded within it. The role of these embedded genes in NF1, if any, remains to be determined. Initial patient mutations have indicated that inactivation of the gene is the mechanism of the disease, implying that the *NF1* gene is likely to be a tumor suppressor. The *NF2* gene, via indirect evidence such as LOH studies, also fits a tumor-suppressor model. The study of the *NF1* gene and cloning and analysis of the *NF2* gene may reveal much about the normal action of tumor suppressors (growth regulators) and involvement of these two genes in cancer progression.

Thus far, molecular genetics has provided the means for prenatal and preclinical diagnosis in NF1 or NF2 families with sufficient pedigree structure by use of linked markers. Now that the *NF1* gene has been identified, diagnosis can be extended to new-mutation patients, although the hunt for a mutation in this very large gene is far from trivial. Homology of the *NF1* gene product with GAP proteins is providing insight into potential functions. Ultimately, research will be aimed at understanding the mechanisms of these genes and their proteins to allow future development of treatments and cures.

SELECTED REFERENCES

Andersen LB, Tommerup N, Koch J. Formation of a minichromosome by excision of the proximal region of 17q in a patient with von Recklinghausen neurofibromatosis. Cytogenet Cell Genet 1990;53:206–210.

Bader JL. Neurofibromatosis and cancer. Ann NY Acad Sci 1986;486:57–65.

Barker D, Wright E, Nguyen K, et al. Gene for von Recklinghausen neurofibromatosis is in the pericentromeric region of chromosome 17. Science 1987; 236:1100–1102.

Buchberg AM, Cleveland LS, Jenkins NA, Copeland NG. Sequence homology shared by neurofibromatosis type 1 gene and *IRA-1* and *IRA-2* negative regulators of the *RAS* cycle AMP pathway. Nature 1990;347:291–294.

Call KM, Glaser T, Ito CY, et al. Isolation and characterization of a zinc finger polypeptide gene at the human chromosome 11 Wilms' tumor locus. Cell 1990; 60:509–520.

Cawthon RM, Andersen LB, Buchberg AM, et al. cDNA sequence and genomic structure of *EVI2B,* a gene lying within an intron of the neurofibromatosis type 1 gene. Genomics 1990;9:446–460.

Cawthon RM, O'Connell P, Buchberg AM, et al. Identification and characterization of transcripts from the neurofibromatosis 1 region: the sequence and genomic structure of *EVI2* and mapping of other transcripts. Genomics 1990; 7:555–565.

Cawthon RM, Weiss R, Xu G, et al. A major segment of the neurofibromatosis type 1 gene: cDNA sequence, genomic structure, and point mutations. Cell 1990;62:193–201.

Cohen MM. Further diagnostic thought about the Elephant Man. Am J Med Genet 1988;29:777–782.

Colley A, Clayton-Smith J, Donnai D. Noonan syndrome and neurofibromatosis type 1 segregating independently in a family. Am J Hum Genet 1989;45:A43.

Collins FS, O'Connell P, Ponder BAJ, Seizinger BR. Progress towards identifying the neurofibromatosis (*NF1*) gene. Trends Genet 1989;5:217–221.

Crowe FW, Schull WT, Neel JF. A clinical, pathological, and genetic study of multiple neurofibromatosis. Springfield, IL: Charles C. Thomas, 1956.

Fialkow PJ, Sagebiel RW, Gartler SM, Rimoin DL. Multiple cell origin of hereditary neurofibromatosis. N Engl J Med 1971;284:298–300.

Fountain JW, Lockwood WK, Collins FS. Transfection of primary human skin fibroblasts by electroporation. Gene 1988;68:167–172.

Fountain JW, Wallace MR, Bruce MA, et al. Physical mapping of a translocation breakpoint in neurofibromatosis. Science 1989;244:1085–1087.

Friend SH, Bernards R, Rogelj S, et al. A human DNA segment with properties of the gene that predisposes to retinoblastoma and osteosarcoma. Nature 1986;323;643–646.

Glover TW, Stein CK, Legius E, Andersen LB, Brereton A, Johnson S. Molecular and cytogenetic analysis of tumors in von Recklinghausen neurofibromatosis. Genes Chromosomes Cancer 1991;3:62–70.

Goldgar DE, Green P, Parry DM, Mulvihill JJ. Multipoint linkage analysis in neurofibromatosis type 1: an international collaboration. Am J Hum Genet 1989;44:6–12.

Harper PS. Gene mapping and neurogenetics. J Med Genet 1987;24:513–514.

Huson SM. Recent developments in the diagnosis and management of neurofibromatosis. Arch Dis Child 1989;64:745–749.

Jadayel D, Fain P, Upadhyaya M, et al. Paternal origin of new mutations in von Recklinghausen neurofibromatosis. Nature 1990;343:558–559.

Knudson AG Jr. Mutation and cancer: statistical study of retinoblastoma. Proc Natl Acad Sci USA 1971;68:820–823.

Leach RJ, Thayer MJ, Schafer AJ, Fournier REK. Physical mapping of human chromosome 17 using fragment-containing microcell hybrids. Genomics 1989;5:167–176.

Ledbetter DH, Rich DC, O'Connell P, Leppert M, Carey JC. Precise localization of *NF1* to 17q11.2 by balanced translocation. Am J Hum Genet 1989;44:20–24.

Levinson B, Kenwrick S, Lakich D, Hammonds G, Gitschier J. A transcribed gene in an intron of the human factor VIII gene. Genomics 1990;7:1–11.

Martuza RL, Eldridge R. Neurofibromatosis 2. N Engl J Med 1988;318:684–688.

Menon AG, Anderson KM, Riccardi VM, et al. Chromosome 17p deletions and p53 gene mutations associated with the formation of malignant neurofibrosarcomas in von Recklinghausen neurofibromatosis. Proc Natl Acad Sci USA 1990;87:5435–5439.

Menon AG, Ledbetter DH, Rich DC, et al. Characterization of a translocation within the von Recklinghausen neurofibromatosis region of chromosome 17. Genomics 1989;5:245–249.

Mikol DD, Gulcher JR, Stefansson K. The oligodendrocyte-myelin glycoprotein belongs to a distinct family of proteins and contains the HNK-1 carbohydrate. J Cell Biol 1990;110:471–480.

Monaco AP, Neve RL, Colletti-Feener, Bertelson CJ, Kurnit DM, Kunkel LM. Isolation of candidate cDNAs for portions of the Duchenne muscular dystrophy gene. Nature 1986;323:646–650.

Mulvihill JJ, moderator. Neurofibromatosis 1 (Recklinghausen disease) and neurofibromatosis 2 (bilateral acoustic neurofibromatosis): an update. Ann Intern Med 1990;113:39–52.

Nigro JM, Baker SJ, Preisinger AC, et al. Mutations in the *p53* gene occur in diverse human tumour types. Nature 1989;342:705–708.

O'Connell P, Leach R, Cawthon RM, et al. Two *NF1* translocations map within a 600-kilobase segment of 17q11.2. Science 1989;244:1087–1088.

O'Connell P, Viskochil D, Buchberg AM, et al. The human homolog of murine *evi2* lies between two von Recklinghausen neurofibromatosis translocations. Genomics 1990;7:547–554.

Riccardi VM. Von Recklinghausen neurofibromatosis. N Engl J Med 1981; 305:1617–1627.

Riccardi VM, Eichner JE. Neurofibromatosis: phenotype, natural history, and pathogenesis. Baltimore: Johns Hopkins University Press, 1986.

Riccardi VM, Lewis RA. Penetrance of von Recklinghausen neurofibromatosis: a distinction between predecessors and descendents. Am J Hum Genet 1988;42:284–289.

Rouleau GA, Seizinger BR, Wertelecki W, et al. Flanking markers bracket the neurofibromatosis type 2 (*NF2*) gene on chromosome 22. Am J Hum Genet 1990;46:323–328.

Rouleau GA, Wertelecki W, Haines JL, et al. Genetic linkage of bilateral acoustic neurofibromatosis to a DNA marker on chromosome 22. Nature 1987; 329:246–248.

Seizinger BR, Rouleau GA, Ozelius LJ, et al. Genetic linkage of von Recklinghausen neurofibromatosis to the nerve growth factor receptor gene. Cell 1987;49:589–594.

Seizinger BR, Rouleau G, Ozelius LJ, et al. Common pathogenetic mechanism for three tumor types in bilateral acoustic neurofibromatosis. Science 1987; 236:317–319.

Skuse GR, Kosciolek BA, Rowley PT. Molecular genetic analysis of tumors in von Recklinghausen neurofibromatosis: loss of heterozygosity for chromosome 17. Genes, Chromosomes Cancer 1989;1:36–41.

Stumpf DA, Alksne JF, Annegers JF, et al. NIH consensus development conference statement. Neurofibromatosis 1987;6.

Sørenson SA, Mulvihill JJ, Nielsen A. Long-term follow-up of von Recklinghausen neurofibromatosis. Survival and malignant neoplasms. N Engl J Med 1986;314:1010–1015.

Tanaka K, Nakafuku M, Satoh T, et al. S. cerevisiae genes *IRA1* and *IRA2* encode proteins that may be functionally equivalent to mammalian *ras* GTPase activating protein. Cell 1990;60:803–807.

Vance JM, Nicholson GA, Yamaoka LH, et al. Linkage of Charcot-Marie-Tooth neuropathy type 1a to chromosome 17. Exp Neurol 1989;104:186–189.

van Tuinen P, Rich DC, Summers KM, Ledbetter DH. Regional mapping panel for human chromosome 17: application to neurofibromatosis type 1. Genomics 1987;1:374–381.

Viskochil D, Buchberg AM, Xu G, et al. Deletions and a translocation interrupt a cloned gene at the neurofibromatosis type 1 locus. Cell 1990;62:187–192.

Wallace MR, Fountain JW, Brereton AM, Collins FS. Direct construction of a chromosome-specific NotI linking library from flow-sorted chromosomes. Nucl Acids Res 1988;17:1665–1677.

Wallace MR, Marchuk DA, Andersen LB, et al. Type 1 neurofibromatosis gene: identification of a large transcript disrupted in three NF1 patients. Science 1990;249:181–186.

Wertelecki W, Rouleau GA, Superneau DW, et al. Neurofibromatosis 2: clinical and DNA linkage studies of a large kindred. N Engl J Med 1988;319:279–283.

Xu G, O'Connell P, Viskochil D, et al. The neurofibromatosis type 1 gene encodes a protein related to GAP. Cell 1990;62:599–608.

CHAPTER 8

Phenylketonuria

Randy C. Eisensmith
Savio L. C. Woo

Phenylketonuria (PKU) and hyperphenylalaninemia (HPA) are autosomal recessive genetic disorders secondary to a deficiency of hepatic phenylalanine hydroxylase. Both disorders are characterized by an accumulation of phenylalanine in the serum, resulting initially in hyperphenylalaninemia but soon after associated with other abnormalities in aromatic amino acid metabolism. Although untreated PKU patients develop severe postnatal brain damage and irreversible mental retardation, the various degrees of mental impairment experienced by these patients can be greatly ameliorated through the rigorous implementation of a low-phenylalanine diet.

PKU is among the most common inborn errors of amino acid metabolism in man, with an average incidence of approximately 1 in 10,000 Caucasian births. With the introduction of mass neonatal screening, a broad spectrum of clinical and biochemical phenotypes has been observed in HPA and PKU patients, ranging from the mildly elevated levels of serum phenylalanine and near-normal intelligence observed in untreated patients with benign HPA to the significantly elevated serum phenylalanine levels and drastically reduced IQ scores observed in untreated patients with classical PKU. This phenotypic heterogeneity may reflect an underlying heterogeneity at the molecular level.

HISTORICAL AND CLINICAL DESCRIPTION

Discovery of Phenylketonuria

Classical phenylketonuria was first described by Folling, who in 1934 reported increased levels of phenylpyruvate in the urine of a specific subgroup of mentally retarded patients. The presence of similar symptoms in several pairs of siblings led Folling to conclude that PKU was probably inherited as an autosomal recessive trait. A year after Folling's discoveries, Penrose confirmed that PKU is an autosomally transmitted recessive genetic disorder.

Although phenylalanine metabolism was immediately implicated in the pathophysiology of PKU, the precise biochemical defect present in PKU patients was not established until 1947, when Jervis observed that the administration of phenylalanine produced a prompt elevation in serum tyrosine levels in normal individuals, but not in individuals with PKU. Furthermore, Jervis subsequently demonstrated the conversion of phenylalanine to tyrosine in postmortem liver samples from normal individuals but not in those from PKU patients. From these studies, Jervis concluded that PKU patients lacked the ability to convert phenylalanine to tyrosine.

Introduction of Dietary Therapy and Newborn Screening

Inspired by the successful use of dietary management in the treatment of galactosemia, Louis Woolf, Horst Bickel, and others soon began to examine the effects of low-phenylalanine diets on various biochemical and clinical indices in PKU patients. By the mid-1950s, several reports appeared describing significant reductions in serum phenylalanine and urinary phenylpyruvate levels in young PKU patients following dietary restriction. In addition, more recent reports have noted slight improvements in the behavioral performance of PKU patients receiving dietary therapy.

One of the most important aspects of these initial studies was the observation of an inverse relationship between the age of onset of dietary therapy and the ultimate IQ level attained by treated patients. In other words, the most serious effects of PKU could be greatly reduced by dietary therapy, but only if affected individuals could be identified and treated early in the neonatal period. This observation provided a strong incentive for the implementation of neonatal screening programs for PKU. With the development of the Guthrie test, an inexpensive, semiquantitative bacterial inhibition assay for the determination of serum phenylalanine levels, such PKU screening programs became practical. From the collective results of mass screening programs in Western countries, the incidence of PKU has been estimated at approximately 1 in 10,000. This value corresponds to a carrier frequency of approximately 1 in 50 for this autosomally transmitted recessive disorder.

Elucidation of the Phenylalanine Hydroxylating System in Man

When Jervis first detected the conversion of phenylalanine to tyrosine in livers of normal individuals, relatively little was known about the enzyme or enzymes responsible for this reaction. By the early 1970s, however, most of the basic factors that influence hepatic phenylalanine hydroxylation has been identified, in large part due to studies performed in the laboratory of Seymour Kaufman. The model for hepatic phenylalanine hydroxylation derived from these studies is summarized in Figure 8.1.

Phenylalanine hydroxylase (PAH) mediates the conversion of the essential amino acid L-phenylalanine to L-tyrosine. In man, PAH is expressed exclusively in the liver. PAH is a mixed-function oxygenase, simultaneously catalyzing both the oxidation of L-phenylalanine and the reduction of molecular oxygen. As shown in Figure 8.1, one of the oxygen atoms in the O_2 molecule is used in the hydroxylation of L-phenylalanine, while the other oxygen atom is reduced to H_2O. Also required in this reaction is the pterin

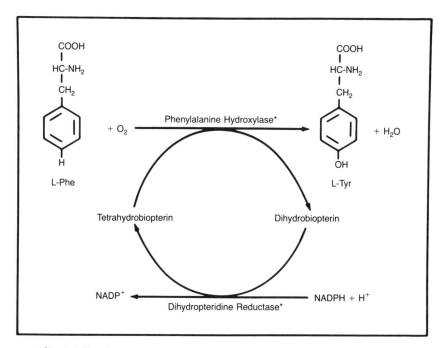

*Classical Phenylketonuria: deficiency of phenylalanine hydroxylase
*Hyperphenylalaninemia: reduced levels of phenylalanine hydroxylase
*Atypical Phenylketonuria: deficiency of dihydropteridine reductase

Figure 8.1: The phenylalanine hydroxylating system in man. The asterisks denote genetic disorders of phenylalanine hydroxylation associated with these two gene products. *Reprinted in part with permission from Eisensmith RC, Woo SLC. Phenylketonuria, molecular genetics. In: The encyclopedia of human biology. New York: Academic Press, 1991.*

cofactor L-erythrotetrahydrobiopterin (BH_4), which serves as an electron donor. The cofactor is oxidized to the corresponding dihydro form during each reaction cycle. Adequate cofactor levels are maintained in vivo through the reduction of dihydrobiopterin to tetrahydrobiopterin by the enzyme quinonoid dihydrobiopterin reductase (QDPR). The characterization of this multienzyme phenylalanine hydroxylating system marked the end of the classical era of PKU research.

Clinical Considerations in PKU

The primary symptom observed in classical PKU patients is mental retardation. Detailed examination of postmortem brain samples from PKU patients have most notably revealed significant alterations in both myelination and protein synthesis. Since the pathophysiology of PKU is poorly understood, it is not yet clear which, if any, of these effects is responsible for the mental impairment seen in PKU patients. Other symptoms commonly reported in PKU patients include the peculiar "mousy" odor that stimulated Folling's initial investigations, hypopigmentation and other dermatological conditions, behavioral disturbances, and convulsive seizures. Additional clinical signs observed in PKU patients include abnormal EEG patterns, impaired postnatal growth affecting head size, and reduced sperm count and semen volume in males.

Although most cases of PKU are associated with various degrees of PAH deficiency, with associated phenotypes ranging from classical PKU to mild PKU or HPA, deficits in enzymes involved in the biosynthesis and metabolism of the cofactor BH_4 can also produce elevations in serum phenylalanine levels that lead to PKU. These "atypical" or BH_4-deficient forms of PKU represent only about 1–2% of all PKU cases. The existence of "atypical" PKU, however, does complicate somewhat the diagnosis of PKU in newborns with elevated serum phenylalanine levels. Furthermore, patients with atypical PKU require different regimens of treatment and counseling and may face significantly different prognoses than patients with PAH deficiencies. Since the biochemical and genetic bases of these two forms of PKU are so fundamentally different, and since "atypical" PKU is rather rare, this discussion will be limited primarily to PAH-deficient disease states.

MOLECULAR GENETICS

Gene Description

Isolation of Human PAH cDNA Clones
Prior to the widespread use of molecular biological techniques, little was known concerning the structure

and location of the gene(s) encoding phenylalanine hydroxylase in man. Preliminary electrophoretic analyses of the human PAH protein suggested that PAH is a polymeric enzyme composed of multiple subunits. Thus, it was not initially clear whether human PAH is a heteropolymer or a homopolymer. This issue was critical, for if the enzyme is a heteropolymer, multiple genetic loci could be expected, further complicating genetic analyses. Preliminary linkage analyses also failed to provide definitive information concerning the nature of the *PKU* locus in man. This situation was quickly improved through the application of molecular methodologies.

The purification of rat phenylalanine hydroxylase and preparation of monospecific antibodies to this enzyme permitted the isolation of rat PAH mRNA from rat liver by polysome immunoprecipitation and the subsequent cloning of the cDNA. The authenticity of this clone was initially established through hybrid-selected translation of the cDNA and confirmed by alignment of the amino acid sequence predicted from the cDNA with that obtained through partial sequencing of the purified rat protein. Using the rat cDNA as a specific hybridization probe, several human cDNA clones were isolated from a human liver cDNA library, the longest of which contained an open reading frame encoding a protein comprised of 451 amino acids. The amino acid sequence of human PAH predicted from the cDNA was shown to be over 90% homologous to the amino acid sequence of the rat enzyme.

Localization of the Human *PAH* Gene   Introduction of an expression vector containing the full-length human PAH cDNA into cultured mammalian cells results in the expression of PAH mRNA and the production of both PAH immunoreactivity and pterin-dependent enzymatic activity similar to that produced by PAH obtained from human liver. Thus, the human PAH cDNA contains all of the information required for the production of functional PAH activity, indicating that the PAH protein is the product of a single gene. This result permits the use of the human PAH cDNA as a probe to perform molecular analysis of the *PAH* gene/*PKU* locus in man.

As alluded to above, preliminary linkage studies using other polymorphic protein markers have generally failed to identify the *PKU* locus in man. Originally, this locus was assigned to chromosome 1 on the basis of data that suggested moderate linkage with the phosphoglucomutase locus *PGM1* and the amylase loci *AMY1* and *AMY2*. However, more extensive studies failed to confirm this result. Using the human cDNA as a hybridization probe, the chromosomal assignment of the human *PAH* gene could be made simply by probing a set of human/rodent cell hybrids that contained different combinations of human and rodent chromosomes. The results of such experiments localized the *PAH* gene to chromosome 12. Subsequent experiments using both deletion chromosome mapping and in situ hybridization not only confirmed this result but also further localized the human *PAH* gene/*PKU* locus to the q22–q24.1 region of chromosome 12.

Structure of the Human *PAH* Gene   The human PAH cDNA was next used as a hybridization probe in Southern analyses of human genomic DNA. These studies indicated that the chromosomal *PAH* gene was greater than 65 kb in length and contained multiple intervening sequences. Due to the large size of this gene, a human genomic DNA library was constructed using the cosmid vector pCV107, and four overlapping cosmid clones were isolated from this library. Detailed analysis of these clones indicated that the human PAH gene is approximately 90 kb long and contains 13 exonic regions separated by introns ranging from 1 to 23 kb in length. Although the human *PAH* gene is large, the mature messenger RNA produced by this gene is only 2.4 kb. Thus, the ratio of noncoding to coding sequences of this gene is among the highest found in eukaryotic genes.

Restriction Fragment–Length Polymorphisms at the Human *PAH* Locus   Using the human PAH cDNA as a specific hybridization probe, Southern analysis was performed on both normal individuals and PKU patients. This analysis revealed the presence of several restriction fragment–length polymorphisms (RFLPs) in or near the human PAH gene. These RFLPs are the result of nucleotide substitutions within the genomic DNA that generate or abolish recognition sites for specific restriction endonucleases. In most cases, these nucleotide substitutions are benign, i.e., they do not alter the normal phenotype. However, in a few cases, RFLP sites lie within the coding region of the gene, and polymorphisms at these sites could produce missense mutations that cause PKU. The relationship of eight RFLP sites to the 13 exons that constitute the complete coding region of the 90-kb human *PAH* gene are shown in Figure 8.2. The four most common of the 70 or so haplotypes identified thus far are shown at the bottom of this figure.

Tissue Specificity/Gene Regulation

Although several early reports described the expression of PAH activity in nonhepatic tissues in man, these observations have not been confirmed in subsequent studies. One recent study has demonstrated the presence of PAH mRNA in leukocytes. Since leukocytes apparently lack detectable amounts of pterin-dependent PAH enzyme activity, the significance of this finding is unclear. Furthermore, as the structure of the PAH promoter has only recently been described, relatively little is currently known about the

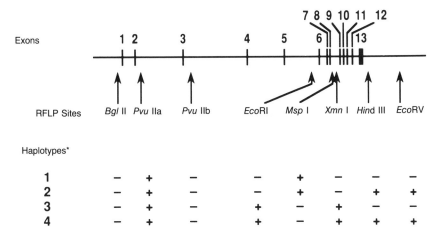

Figure 8.2: RFLP haplotypes at the human *PAH* locus. The exonic structure in relation to the eight most frequent polymorphic sites of the human *PAH* gene (top). The four most prevalent PAH haplotypes in the Northern European population (bottom). The pluses and minuses denote the presence or absence of a particular-sized fragment (allele) following digestion of genomic DNA with a given restriction endonuclease. *Reprinted with permission from Eisensmith RC, Woo SLC. Phenylketonuria, molecular genetics. In: The encyclopedia of human biology. New York: Academic Press, 1991.*

mechanisms that may account for the observed tissue-specific expression of PAH in man.

Genetic Lesions Accounting for PAH Deficiencies

Haplotype analysis performed on PKU kindreds collected from several European countries demonstrated the predominance of two particular PAH haplotypes (haplotypes 2 and 3) among mutant chromosomes. Since RFLP haplotypes and specific mutations had previously been shown to be tightly linked in genetic disorders such as β-thalassemia, haplotype 2 and 3 PKU chromosomes were next examined by direct molecular analysis.

Molecular Analysis of Haplotype 2 and 3 PKU Chromosomes The *PAH* genes from individuals bearing mutant PAH haplotype 2 or 3 chromosomes were isolated by molecular cloning, and the coding region was examined by sequence analysis. These analyses revealed the presence of a C-to-T transition in exon 12 of the *PAH* gene from an individual homozygous for mutant haplotype 2. This missense mutation results in the substitution of tryptophan for arginine at residue 408 of the PAH protein (the "R408W" mutation). Similar analyses revealed the presence of a G-to-A transition at the consensus splice–donor site at the exon 12/intron 12 boundary region in an individual homozygous for mutant haplotype 3 (the "splicing" mutation).

In both cases, expression analysis was performed to verify that these single-base substitutions were in fact the cause of PKU. For the R408W mutation, site-directed mutagenesis was used to create an expression vector containing the mutagenized PAH cDNA. Normal and R408W mutant constructs were then introduced into cultured mammalian cells and levels of PAH mRNA, immunoreactive PAH protein and PAH enzyme activity were assayed. In these studies, both the normal and R408W mutant constructs produced similar levels of PAH mRNA, but the R408W mutant construct did not produce detectable PAH enzyme activity or immunoreactivity. Thus, the R408W mutation present on mutant haplotype 2 chromosomes could in fact cause PKU.

The effects of the "splicing" mutation were also examined in expression studies. In this case, an expression vector containing a mini-*PAH* gene was constructed by inserting a specific restriction fragment containing the intron 12 "splicing" mutation into the appropriate site of the PAH cDNA. The results of these studies were essentially similar to those seen previously with the R408W mutant construct. No detectable amounts of PAH enzyme activity or immunoreactivity were observed in these studies, although aberrant mRNA species were observed.

Linkage Disequilibrium between Specific Mutations and PAH Haplotype 2 and 3 Mutant Chromosomes   The precise identification of the molecular lesions present on mutant haplotype 2 and 3 chromosomes was quickly followed by direct detection of these mutations in genomic DNA using allele-specific oligonucleotide (ASO) hybridization analysis. Such analyses have demonstrated the presence of the R408W mutation on nearly all haplotype 2 mutant chromosomes and the "splicing" mutation on nearly all haplotype 3 mutant chromosomes in Northern European populations. The R408W mutation was not observed on the vast majority of non–haplotype 2 mutant chromosomes in most European populations and has never been observed on normal chromosomes. These results confirm that the R408W mutation is responsible for PKU and suggest that it is distinct from other mutations that are present on other mutant chromosomes. Similar results have been observed for the "splicing" mutation and haplotype 3 mutant chromosomes. The absence of these two mutations from the majority on non–haplotype 2 or 3 PKU chromosomes implies that these other PKU chromosomes must bear novel PKU mutations, further confirming the heterogeneous nature of this disorder at the molecular level.

The close association observed between haplotype 2 mutant alleles and the R408W missense mutation or between haplotype 3 mutant alleles and the "splicing" mutation suggests that these mutational events occurred relatively recently on normal haplotype 2 or normal haplotype 3 chromosomes. Although the resulting PKU chromosomes have since been distributed throughout many European populations, fragments bearing these mutations have not yet been transferred to chromosomes of other haplotypes by crossover or gene conversion events in these populations. Thus, these mutations are still in linkage disequilibrium with their particular chromosomal haplotype in most human populations.

Molecular Analysis of PAH Haplotype 1 and 4 Mutant Chromosomes

Sequence and hybridization analyses similar to those described previously have also been performed on mutant haplotype 1 and 4 chromosomes from various European populations. These studies have revealed the presence of at least two missense mutations associated with haplotype 1 PKU chromosomes of Northern Europeans. Both mutations are G-to-A transitions in exon 7 of the *PAH* gene. One mutation results in the substitution of glutamine for arginine at residue 261 of the PAH protein (the "R261Q" mutation), while the other mutation results in the substitution of lysine for glutamate at residue 280 of the PAH protein (the "E280K" mutation).

Similarly, at least two mutations have been observed on haplotype 4 PKU chromosomes in Northern Europeans. A G-to-A transition in exon 5 of the *PAH* gene results in the substitution of glutamine for arginine at residue 158 of the PAH protein (R158Q), while a C-to-T transition in exon 7 results in the conversion of an arginine codon at position 243 to a termination codon (R243X). Population screening studies have demonstrated that, unlike haplotype 2 or 3 mutant chromosomes, many of the haplotype 1 or 4 mutant chromosomes do not bear the same molecular lesion. These results indicate that mutant haplotype 1 and 4 chromosomes bear multiple, independent mutations, at least two of which are in linkage disequilibrium with respect to their haplotype background. The presence of multiple mutations on mutant haplotype 1 and 4 chromosomes may reflect the predominance of these two haplotypes among normal *PAH* alleles.

Molecular/Clinical Heterogeneity

Because of the linkage disequilibrium between the R408W and "splicing" mutations and specific RFLP haplotypes, it is possible, within specific populations, to infer the genotype of PKU patients from the presence of mutant haplotype 2 or 3 chromosomes. Thus, PKU patients who are homozygous for mutant haplotype 2 or 3 most likely bear the same mutation on both chromosomes, while those individuals containing both mutations are most likely compound heterozygotes for chromosomes bearing both of these mutations. Similarly, the tight linkage disequilibrium observed between mutation and haplotype for the R408W mutation or the "splicing" mutation suggests that patients who are heterozygous for mutant haplotype 2 or 3 with any of the other haplotypes are also likely to be compound heterozygotes. By this type of analysis, it is estimated that about 75% of all PKU patients in the Northern European population are genetic compounds. This result further demonstrates the considerable molecular heterogeneity at the *PKU* locus.

A correlation has recently been reported between specific DNA haplotypes and clinical phenotypes in PKU patients. Most patients homozygous or compound heterozygous for haplotype 2 and 3 mutant chromosomes show severe biochemical and clinical phenotypes, whereas most patients who carry mutant alleles of either haplotype 1 or 4 exhibit milder phenotypes. However, since all mutant chromosomes of a given haplotype may not bear the same genetic lesion, this correlation is somewhat limited in its implications regarding the molecular basis for phenotypic diversity in PKU.

With the recent identification of several PKU mutations associated with specific PAH haplotypes in Caucasians and the biochemical characterization of the corresponding mutant proteins, the correlation between mutant haplotypes and clinical phenotypes has been extended to the molecular level. For example, no detectable PAH enzyme activity or immunoreactivity is observed following the in vitro expression of either the R408W or "splicing" mutant constructs, and patients homozygous or compound heterozygous for these mutant alleles manifest severe biochemical and clinical phenotypes. In contrast, in vitro expression analysis of either the R158Q or R261Q mutant constructs demonstrates the production of reduced amounts of PAH enzyme activity and immunoreactivity relative to the normal construct. Since mutations conferring residual PAH activity are generally dominant over mutations that confer no enzyme activity, patients bearing R158Q or R261Q mutant alleles, either as homozygotes or heterozygotes, generally exhibit milder biochemical and clinical phenotypes than patients bearing the R408W or "splicing" mutant alleles. Thus, there appears to be an underlying molecular basis for the phenotypic heterogeneity observed in PKU and HPA patients, suggesting that genotype may ultimately be used to predict the severity of phenotype, to optimize dietary therapy, and to predict prognosis in PKU and HPA patients.

MOLECULAR MEDICINE

Diagnosis of PKU

Traditionally, diagnosis of PKU has been based primarily on the results of newborn screening, using the Guthrie or other more quantitative tests, with subsequent testing to differentiate PAH- or BH_4-deficient disease states. Testing for BH_4-deficient phenotypes involves the intravenous injection of BH_4, which is accompanied by an abrupt decrease in serum phenylalanine levels in patients who cannot normally synthesize the endogenous cofactor. Additional testing must be performed to precisely define other deficiencies related to pterin synthesis and metabolism. Phenylalanine or protein-loading tests have also been employed to diagnose the severity of the PAH deficiency in PKU and HPA patients. In general, these tests are somewhat limited, in that there is a significant degree of overlap between patients with different phenotypes, as well as between carriers and normal individuals.

More recently, various molecular methodologies have been applied for the detection of PKU alleles, both prenatally and in carriers. As mentioned previously, several RFLPs are tightly linked to the *PAH* gene. These RFLPs can thus be used to follow the transmission of normal or mutant alleles in PKU kindreds by comparing the hybridization patterns of parental and proband DNA samples following digestion with one or more restriction endonucleases. Figure 8.3 illustrates the first practical application of RFLP haplotype analysis in the detection of mutant PAH chromosomes in a PKU family. Genomic DNAs isolated from the father, the mother, the proband and an unborn child were digested with the restriction endonuclease *Hin*dIII and analyzed by Southern hybridization using the human PAH cDNA as a probe. Both parents are heterozygous for the *Hin*dIII polymorphism, as evidenced by the presence of both the 4.2 and 4.0 kb bands, while the proband contains only the 4.0 kb band. Thus, the mutant *PAH* gene is associated with the 4.0 kb allele and haplotype analysis is informative in this family. Fetal DNA was then obtained and similarly analyzed. The fetus proved to be homozygous for the 4.0 kb fragment, permitting the unequivocal prenatal diagnosis of PKU. This diagnosis was confirmed shortly after birth.

Although the application of haplotype analysis permitted the development of prenatal diagnosis of PKU, this technique is not without limitations. Most importantly, prenatal diagnosis using RFLP analysis can only be provided to those families with a prior incidence of PKU. Unfortunately, the vast majority of new cases of PKU (over 95%) are the result of

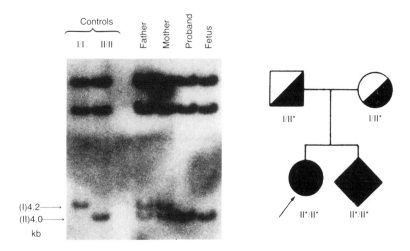

Figure 8.3: RFLP haplotype analysis illustrating Mendelian segregation and inheritance of mutant *PAH* alleles in a family with a prior history of phenylketonuria (see text for details). *Reprinted with permission from Eisensmith RC and Woo SLC. Phenylketonuria, molecular genetics. In: The encyclopedia of human biology. New York: Academic Press, 1991.*

random mating events, and thus are undetectable by traditional haplotype analysis. To overcome these limitations, the specific molecular lesions responsible for PKU must be isolated and characterized, so that genotype can be established by direct molecular analysis. One must remember, however, that PKU diagnosis by genotyping is unlikely to ever be 100% effective, since a negative test does not totally preclude the presence of unknown or novel PKU mutations. Thus, the Guthrie test and other similar biochemical procedures will long remain the standard methods for detecting and diagnosing PKU patients.

Prevention of PKU through Carrier Screening and Prenatal Diagnosis

Since the majority of new PKU mutations are the result of random matings, relatively little can be done to prevent new cases of PKU without methods of reliably detecting carriers in the general population. To this end, carrier screening using a variety of molecular methods have been proposed as a means of reducing the incidence of PKU. If such screening procedures are to be effective, they must be able to detect a significant percentage of PKU mutations. Table 8.1 illustrates the theoretical reduc-

Table 8.1:   Prevention of PKU by Carrier Screening

Screening Accuracy	Carrier Frequency	PKU Frequency	PKU Incidence*
0%	$(\frac{1}{50})^2 \times \frac{1}{4} =$	$\frac{1}{10,000}$	400
50%	$(0.50 \times \frac{1}{50})^2 \times \frac{1}{4} =$	$\frac{1}{40,000}$	100
75%	$(0.25 \times \frac{1}{50})^2 \times \frac{1}{4} =$	$\frac{1}{160,000}$	25
90%	$(0.10 \times \frac{1}{50})^2 \times \frac{1}{4} =$	$\frac{1}{1,000,000}$	4

*Based on 4×10^6 annual births in the United States. *Reprinted in part with permission from Eisensmith RC, Woo SLC. Phenylketonuria, molecular genetics. In: The encyclopedia of human biology. New York: Academic Press, 1991.*

tions in the number of new cases of PKU in the U.S. population that could result from the implementation of carrier screening programs with various levels of accuracy. Currently, in the absence of any carrier screening, approximately 400 new PKU patients are born annually in the United States. However, if carrier screening programs could identify 50% of all carriers of mutant *PAH* alleles in the U.S. population, as is currently the case, and if as a result of carrier screening, matings between carriers did not produce affected children, roughly 300 cases of PKU could be eliminated annually.

As additional PKU mutations are identified, the accuracy level of carrier screening can be increased, resulting in even more dramatic reductions in the incidence of PKU. For example, with the characterization of 90% or more of the *PAH* mutations in the general population, a value not wholly unreasonable for some populations, the frequency of PKU could theoretically be reduced to 1 in 10^6, corresponding to only a few new cases of PKU each year in the United States.

One potential obstacle to the widespread application of carrier screening for PKU mutations by ASO hybridization analysis is the amount of genomic DNA required to produce a detectable hybridization signal. However, this limitation has for the most part been removed through the application of polymerase chain reaction methodologies. Figure 8.4 illustrates the results of carrier screening for the R408W mutation using a combination of PCR and ASO techniques. As shown in this figure, the combined application of these methodologies permits the unequivocal determination of genotype in normal individuals (as in Figure 8.4, Patient A), in carriers (Patient B) and in PKU patients (Patient C). In addition, these methodologies have been refined to such a degree that only a single drop of blood, even dried blood, or a single strand of hair can be used as the source of genomic DNA. Conceivably, these testing procedures could also be fully automated, permitting the rapid analysis of many samples simultaneously. With these techniques, and with the characterization of

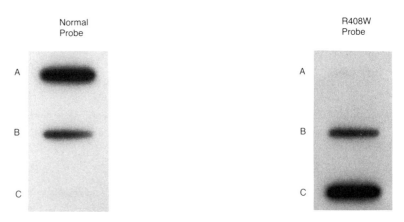

Figure 8.4: PCR amplification and ASO hybridization analysis for the detection of the R408W mutant *PAH* allele in three individuals (A, B, C). A specific mutation-bearing subgenomic fragment was PCR-amplified from the genomic DNA of these three individuals and subjected to ASO hybridization analysis using probes specific for the normal (left) or mutant (right) sequence at this region. Patient A is homozygous normal for this allele. Patient B is heterozygous for the normal and the R408W allele. Patient C is homozygous for the R408W allele.

additional *PAH* mutations, carrier detection of PKU in the general population will soon be technologically feasible.

Before carrier screening for PKU can be implemented in a given population, however, serious consideration must be given to the potential medical, ethical, and social effects of screening. The test should be voluntary and confidentiality is essential. Test results should be used only to make informed reproductive decisions. In addition, the implications of a negative result in a screening test must be adequately addressed. As mentioned previously, the accuracy of carrier detection by gene screening can never be 100%, due to the presence of novel mutations. Thus, a negative result does not totally preclude the possibility of producing an affected offspring. Consequently, the public must also be educated about the relative limitations of such testing procedures, i.e., that screening can reduce but not completely eliminate the risk of developing PKU or similar genetic disorders. Obviously, the magnitude of the reduction in the incidence of the genetic disorder is ultimately dependent upon the percentage of mutant genes that can be reliably detected in a given population. This is in turn determined by analyzing the individuals that comprise these populations.

Therapeutic Options and Directions

The application of molecular methodologies has certainly improved the prospects for successful prenatal diagnosis and carrier screening programs for PKU. Unfortunately, there is still little that can be done for individuals with PKU beyond the administration of dietary therapy. Furthermore, questions remain concerning the appropriate period for cessation of therapy in PKU patients. Some studies have demonstrated significant reductions in IQ scores in affected children following termination of dietary therapy, and thus it has been suggested that treatment should continue at least through the first decade of life. Other more recent studies have suggested that termination of treatment is accompanied by decreases in mental and behavioral performance regardless of the age of the patient at the time therapy was ended. Thus, although therapy can, for the most part, drastically ameliorate the mental impairment experienced by PKU patients, dietary therapy is not without some limitations, specifically regarding compliance and duration of treatment.

The demonstration of PAH activity in cultured mammalian cells following the introduction of a full-length human PAH cDNA raises the possibility that PKU may one day be treated through the introduction of a "replacement gene" into liver cells of affected individuals, thereby enabling them to convert phenylalanine to tyrosine. Unfortunately, the methods of gene transfer employed in these studies are not amenable for the specific transfer of genes with high efficiency. Accordingly, several more efficient and specific methods for the introduction of human PAH into hepatocytes or other cells are currently being examined.

Perhaps the most promising method for the specific introduction of cDNAs into hepatocytes involves the use of recombinant retroviruses. Many features of the retrovirus life cycle make them attractive candidates for gene transfer. First, retroviruses recognize specific features on the surface of target cells, permitting specific targeting of a particular cell type. Second, retroviral infection results in the introduction of the retroviral RNA genome and reverse transcriptase into the target cell cytoplasm. The retroviral RNA is then transcribed into DNA that is randomly inserted into the genomic DNA of the host cell. Thus, infected cells are stably transformed at high efficiency. Third, the proviral DNA is efficiently transcribed and translated by the host cell machinery, resulting in the expression of viral proteins and the production of viral capsules. Finally, the viral genome is inserted or "packaged" into these viral capsules and the mature virus is released from the infected cell.

The key to the design of retroviral-based transfer vectors is the specific packaging sequence present in the retroviral genome. Deletion of the packaging sequence does not interfere with the ability of the virus to produce mature, functional viral capsules, but does prevent the incorporation of the viral genome into these viral capsules. Thus, introduction of an expression vector containing a heterologous cDNA flanked by the appropriate packaging signals into cells previously infected with packaging-defective retrovirus results in the production and release of viral capsules containing the recombinant RNA transcript. These recombinant retroviruses can then be used to specifically infect target cells.

Preliminary experiments using retroviral-based constructs have been performed in hepatoma cells. Hepatoma cells were selected as target cells in these experiments for several important reasons. First, being derived from hepatocytes, hepatoma cells are sufficiently similar to the ultimate target cells. Second, although hepatoma cells do not express PAH, they do contain all of the requisite factors for the reconstitution of the phenylalanine hydroxylating system. Finally, since tyrosine is essential for the survival of hepatoma cells in culture, infected cells can be selected on the basis of their ability to survive in tyrosine-free media.

The successful reconstitution of the phenylalanine hydroxylating system in retroviral-infected cells is demonstrated in Figure 8.5. Hepatoma cells infected with the control virus (left panel) fail to survive in tyrosine-free media, while those infected with the recombinant retrovirus (right panel) thrive under these conditions. This result demonstrates the successful transformation of mammalian cells by recombinant retroviruses, a result that has since been extended to primary hepatocytes.

Initial results of these experiments suggest that retroviral-mediated gene transfer can in fact be used to introduce replacement genes into primary hepatocytes in vitro. However, the success of this approach may be limited by the efficiency with which infected hepatocytes may be reintroduced into the host liver. To monitor the long-term survival of reimplanted hepatocytes, primary hepatocytes from transgenic c57 black mice, which actively secrete the liver-specific protein human α_1-antitrypsin into the serum, were injected into the portal vein of congenic mice, and serum levels of human α_1-antitrypsin were determined at specific intervals. In either case, significant amounts of human α_1-antitrypsin could be detected up to 90 days after injection. Although more detailed studies are still required, the results of these studies certainly suggest that somatic gene therapy through a process of explantation, in vivo infection, and reimplantation, may be possible for the treatment of PKU. Whether or not such therapies will ultimately be implemented for PKU is a question that cannot yet be answered.

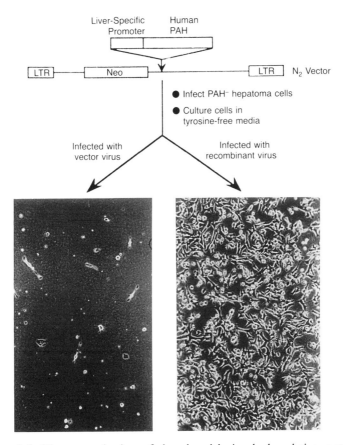

Figure 8.5: The reconstitution of the phenylalanine hydroxylating system in hepatoma cells infected with a PAH-containing retroviral construct (top). Hepatoma cells infected with the control virus (left panel) do not grow in tyrosine-free media, while infected cells (right panel) continue to grow under these conditions. *Reprinted with permission from Scriver CR, Kaufman S, Woo SLC. The hyperphenylalaninemias. In: Scriver CR, Beaudet SL, Sly WS, Valle D, eds. The metabolic basis of inherited disease, volume II. 6th ed. New York: McGraw-Hill, 1989.*

ACKNOWLEDGMENTS

This work was supported in part by National Institutes of Health grant HD-17711 to S.L.C. Woo, who is also an Investigator with the Howard Hughes Medical Institute.

SELECTED REFERENCES

Bickel H, Gerrard J, Hickmans EM. The influence of phenylalanine intake on the chemistry and behavior of a phenylketonuria child. Acta Paediatr Scand 1954;43:64–77.

Güttler F. Hyperphenylalaninemia: diagnosis and classification of the various types of phenylalanine hydroxylase deficiency in childhood. Acta Paediatr Scand 1980;280(suppl):7–80.

Güttler F, Ledley FD, Lidsky AS, DiLella AG, Sullivan SE, Woo SLC. Correlation between polymorphic DNA haplotypes at phenylalanine hydroxylase locus and clinical phenotypes of phenylketonuria. J Pediatr 1987;110:68–71.

Kaufman S. Phenylketonuria and its variants. Adv Neurochem 1976;2:1–132.

Ledley FD, Woo SLC. Prospects for somatic gene therapy of phenylketonuria. In: Kaufman S, ed. Amino acids in health and disease: new perspectives. UCLA symposia on molecular and cellular biology, new series, volume 55. New York: Alan R Liss Inc, 1987.

Lidsky AS, Güttler F, Woo SLC. Prenatal diagnosis of classic phenylketonuria by DNA analysis. Lancet 1985;1:549–551.

Okano Y, Wang T, Eisensmith RC, Güttler F, Woo SLC. Recurrent mutation in the human phenylalanine hydroxylase gene. Am J Hum Genet 1990;46:18–25.

Okano Y, Wang T, Eisensmith RC, Steinmann B, Gitzelmann R, Woo SLC. Missense mutations associated with RFLP haplotypes 1 and 4 of the human phenylalanine hydroxylase gene. Am J Hum Genet 1989;46:18–25.

Robson KJH, Chandra T, MacGillivray RTA, Woo SLC. Polysome immunoprecipitation of phenylalanine hydroxylase mRNA from rat liver and cloning of its cDNA. Proc Natl Acad Sci USA 1982;79:4701–4705.

Sarkar G, Sommer SS. Access to a messenger RNA sequence or its protein product is not limited by tissue or species specificity. Science 1989;244:331–334.

Scriver CR, Kaufman S, Woo SLC. The hyperphenylalaninemias. In: Scriver CR, Beaudet AL, Sly WS, Valle D, eds. The metabolic basis of inherited disease, volume II. 6th ed. New York: McGraw-Hill Inc, 1989.

Wang T, Okano Y, Eisensmith RC, et al. Molecular genetics of PKU in Eastern Europe: a nonsense mutation associated with haplotype 4 of the phenylalanine hydroxylase gene. Somat Cell Mol Genet 1990;16:85–89.

Woo SLC. Molecular basis and population genetics of phenylketonuria. Biochemistry 1989;28:1–7.

Woolf LI, Griffiths R, Moncrieff A. Treatment of phenylketonuria with a diet low in phenylalanine. Br Med J 1955;1:57–64.

CHAPTER 9

Gilles de la Tourette Syndrome

Michele C. LaBuda
David L. Pauls

The classic paper of Gilles de la Tourette provides a comprehensive overview of the syndrome of multiple motor and vocal tics that now bears his name. The stability of this description over the century and more since its initial publication is a credit to the clinical skills of Gilles de la Tourette.

In his report, Gilles de la Tourette described an apparently inherited disorder, affecting individuals of otherwise normal mental and physical health, which was observed to begin in childhood or adolescence and affect males more frequently than females. The disorder was said to be characterized by motor incoordination or movements generally beginning in the face or upper extremities and eventually involving the lower extremities. The movements were noted to be unique due to the abruptness and rapidity with which they were executed. The motor tics were accompanied by vocal tics such as simple noises or the repetition of words. In addition, Gilles de la Tourette observed that during the course of the syndrome, an individual would experience a waxing and waning of symptoms, with new symptoms replacing old ones, and periods of symptom exacerbation, suppression, and nearly total, although temporary, remission.

Although for many decades, Tourette's syndrome (TS) research was sparse, the 1960s brought about renewed interest in the disorder when the therapeutic value of the drug haloperidol was discovered. As a result, many investigations concerning the pathophysiology of TS were initiated. The past decade has brought about a wealth of research regarding such aspects of TS as phenomenology, pathophysiology, and treatment. In contrast to the majority of the disorders discussed in the present volume, however, little can be said with certainty regarding the etiology of TS. Therefore, the purpose of the present chapter is to describe the natural history of TS, to provide the etiological and pathobiological evidence that indicates that TS results from an inherited neurological vulnerability, and to discuss various treatment strategies.

DIAGNOSIS AND NATURAL HISTORY

The diagnostic criteria for TS as established by the Diagnostic and Statistical Manual of Mental Disorders include sudden, involuntary, repetitive motor movements and vocal productions, which do not have to be expressed concurrently but must have begun before 21 years of age. In addition, the tic symptoms must have been present for a minimum of one year and the type of tic or its frequency, complexity, or severity must change over time.

TS is part of a spectrum of tic disorders including chronic motor or vocal tics, transient tics, and tics not otherwise specified. Although the criteria for chronic tic disorder are similar to those for TS, TS is differentiated from chronic tics by the presence of *both* motor and vocal tics. In general, the severity of symptoms and subsequent functional impairment are lesser in magnitude for chronic tics as opposed to TS. Single or multiple motor and/or vocal tics that fulfill the criteria for TS except for having a duration of less than 12 consecutive months, are termed transient tics. The diagnosis of tic disorders not otherwise specified is reserved for tic behaviors not meeting the criteria for any previous diagnostic classification (e.g., tics beginning in adulthood). As will be discussed in a subsequent section, there is evidence to suggest that tic disorders (particularly chronic tics and TS) may constitute a continuum of dysfunction, with TS representing the greatest degree of severity.

Prevalence and Gender Differences

Until recently, no population-based estimates of the prevalence of TS were available. Estimates based on treatment populations varied considerably but all indicated that TS was a relatively rare disorder. A major geographical survey recently completed in North Dakota estimated the prevalence of TS in individuals 18 years of age or younger to be 9.3 and 1.0 per 10,000 for males and females, respectively. The prevalence rates for adults were .77 and .22 per 10,000 for males and females, respectively. A county-wide survey in Rochester, Minnesota estimated that 1,000 new cases of TS would be diagnosed each year in the United States. Whereas these two surveys used cases previously identified due to treatment in order to estimate prevalence, a survey in Monroe County, New York attempted to identify both treated and untreated cases of TS. A highly similar prevalence estimate for TS of 2.87 per 10,000 was obtained. It should be noted that

TS has not been found to be differentially prevalent in any ethnic group or social class.

From the preceding prevalence estimates, it can be seen that males are affected with TS more frequently than are females. From the early family history studies, it appeared that this gender difference could be explained by the existence of a differential threshold of liability, with males being at greater risk. This hypothesis predicts that relatives of affected females are at higher risk for developing TS than are relatives of affected males. In contrast, a recent family study based upon direct interviews of relatives indicates that relatives of affected males and females may be at equal risk. As will be discussed in a subsequent section, results from recent studies suggest that the observed gender difference may be the result of sex-specific expression of symptoms.

Progression and Variability of Symptoms

As shown in Table 9.1, individuals affected with TS exhibit a variety of motor and phonic symptoms. The progression of motor symptoms is often suggested to be cephalocaudal due to the later age of onset of tics involving the trunk and legs than those involving the eyes, face, and head. Whether or not this progression holds true at the level of the individual, however, remains questionable. Nevertheless, motor tics precede vocal tics by approximately two years on average.

Intriguing contributions to the variability of tic expression are occasions of symptom exacerbation and suppressibility. For example, affected individuals have frequently reported an increase in the number of tics in situations of emotional stress. On the other hand, a significant decrease in tic behavior may be observed for a finite period of time (e.g., during a visit to the physician's office, or a time of intense, pleasurable concentration). Although originally thought to the contrary, recent studies have revealed the occurrence of tics during sleep.

Associated Symptoms

Several clinical symptomatologies are reportedly found to be associated with TS, including learning and attentional difficulties, hyperactivity, depressive reactions, and obsessive-compulsive behaviors. Although estimates vary widely from study to study, these types of behaviors are generally found to occur more frequently in individuals suffering from TS than in the general population. As will be detailed in a subsequent section, such findings have raised the question of whether these disorders share a common underlying etiological process.

Table 9.1:   Average Age of Symptom Onset in TS Patients (*N* = 75)

Symptom	Percent Ever Experiencing Symptom	Approximate Average Age of Onset (Yrs)
Motor Tics		
Eyes, face, head	94.7	7.2
Shoulder, neck	92.0	8.7
Arms, hands	82.7	9.1
Trunk	61.3	9.5
Legs	61.3	10.1
Vocal Tics		
Low noises	84.0	10.1
Loud noises	66.7	10.1
Stuttering	26.7	10.5
Repetition	54.7	11.0
Obscenities	37.3	11.0
Syllables	26.7	11.3
Blocking	37.3	11.7
Words out of context	29.3	12.2
Echolalia	37.3	12.7
Compulsive Actions		
Head banging	17.3	9.1
Kissing	14.7	9.2
Touching objects	54.7	9.8
Kicking	36.0	10.1
Tapping	42.7	10.9
Touching of self/others	52.0	11.1
Biting self	12.0	11.7
Touching sexual organs	36.0	11.8
Mimicking others physically	21.3	11.8

Adapted from Jagger J, Prusoff BA, Cohen DJ, Kidd KK, Carbonari CM, John K. The epidemiology of Tourette's syndrome: a pilot study. Schiz Bull 1982;8:267–278.

Prognosis

From longitudinal and cross-sectional studies of TS adults, it appears that, for many individuals, symptoms are at their worst during the first decade after onset. After reaching maturity, however, many affected individuals experience improvements that steadily increase throughout their adult years. With regard to social adjustment in adolescence, it appears that children with TS are at significant risk for poor peer relationships as assessed by classmates' ratings. With regard to life adjustment, results from a large survey conducted in Ohio indicate that although TS adults experience a significantly higher rate of unemployment than average, rates of marriage and divorce are not different from those observed in the general population.

ETIOLOGY

Environmental Influences

A recent twin study of TS reported a concordance rate of 53% among monozygotic twin pairs. Another 24% could be considered concordant if the criteria were broadened to include chronic tic disorders along with TS. Even so, in 23% of these genetically identical twin pairs only one member was affected. This indicates the importance of environmental factors in the etiology of TS. In fact, further analysis indicated that in this sample of discordant identical twin pairs, the birth weights of the affected twins were significantly lower than those of their unaffected co-twins. An earlier study found that mothers of nontwin children with tics were one and a half times more likely to have experienced a complication during pregnancy than were mothers of nontwin children without tics. Although neither study specifically identifies potential prenatal risk factors, they provide interesting preliminary evidence to suggest the importance of environmental influences in the etiology of TS.

Genetic Influences

Twin and Family Studies Results from the twin study previously discussed showed a concordance rate for TS of 8% in the dizygotic twin pairs. Because monozygotic twins (MZ) are genetically identical, whereas dizygotic twins (DZ) share only half their genes on average, the greater MZ than DZ concordance rate is indicative of genetic involvement in the etiology of TS.

Similarly, family history studies indicate that relatives of TS individuals are at greater risk for developing TS than are individuals in the general population. Information obtained by direct interview methods confirms the finding of increased risk in relatives of TS patients. Estimates of the risk to first-degree relatives range from 15 to 30%.

Family studies have also been informative with respect to the relationship between TS and other disorders. For example, using the rates of TS and chronic tics in the relatives of affected individuals or probands, Pauls et al. showed the two disorders to be etiologically related, with chronic tics being a milder manifestation of TS. Data from subsequent family studies were used to test whether TS shared a common etiology with either attention deficit disorder (ADD) or obsessive-compulsive disorder (OCD). As

Table 9.2: Frequency of Tourette Syndrome (TS), Chronic Tics, Attention Deficit Disorder (ADD), and Obsessive-Compulsive Disorder (OCD) in Relatives of TS Probands

	Diagnosis in Proband			Diagnosis in Proband	
Diagnosis in Relatives	TS With ADD	TS Without ADD	Diagnosis in Relatives	TS With OCD	TS Without OCD
TS	10%	11%	TS	12%	9%
Chronic tics	17%	20%	Chronic Tics	17%	20%
ADD	17%	2%	OCD	19%	27%

Adapted from Pauls DL, Towbin KE, Leckman JF, Zahner GEP, Cohen DJ. Gilles de la Tourette's syndrome and obsessive-compulsive disorder: evidence supporting a genetic relationship. Arch Gen Psychiatry 1986;43:1180–1182; and Price RA, Pauls DL, Kruger SD, Caine ED. Family data support a dominant major gene for Tourette syndrome. Psychiatric Res 1988;24:251–261.

shown in Table 9.2, the rate of ADD in relatives of probands with TS and ADD was approximately eight times higher than the rate of ADD in relatives of TS probands without concomitant ADD, results that are consistent with the hypothesis of a separate genetic etiology for TS and ADD. In contrast, the rate of OCD in relatives of probands with TS and OCD was comparable to the rate of OCD in relatives of proband with TS but not OCD. These analyses suggest that TS and OCD may be etiologically related.

Segregation Analyses Several groups of researchers have attempted to fit genetic models to family data. Although variability exists between studies, as seen in Table 9.3, the transmission of TS is best explained by an incompletely penetrant, autosomal dominant gene.

Results from a series of segregation analyses reported by Pauls yield further insight into the etiological relationships between TS, chronic tics, and obsessive-compulsive disorder. Although all analyses concluded that an autosomal dominant mode of transmission best fit the data, separate analyses based upon differing diagnostic criteria did not always accurately predict the familial pattern of illness within the families. For example, when affection status was defined as either TS or chronic tics, there appeared to be too many affected fathers and too few affected mothers. When the diagnostic criteria were altered, however, and individuals with OCD were also classified as affected, the fit between the observed and expected numbers of ill family members was comparable. It was argued that these results lend further credence to the notion that TS, chronic tics, and OCD are part of the same genetic disorder and that the observable expression of the TS gene is sex-specific (i.e., males with the TS gene may

Table 9.3: Summary of Published Segregation Analyses of TS

				Results					
						Penetrance*			
Study	N	Genetic Model†	Gene Frequency	Gender	TT	Tt	tt	Phenocopies	
Baron, 1981	123	SML	.3%	Males	.953	.812	.018	Males	79%
				Females	.875	.640	.004	Females	53%
Kidd and Pauls, 1982	118	SML	3%	Males	.500	.023	.000	Males	—
				Females	.106	.001	.000	Females	—
Comings et al., 1984	242	Semi-dominant	.5%	Combined	.940	.470	.003	Combined	35%
Devor, 1984	35	Semi-dominant	4%	Combined	.999	.100	.001	Combined	.6%
Pauls et al., 1986	27	Dominant	.6%	Males	1.000	1.000	.002	Males	1.6%
				Females	.709	.709	.000	Females	0%
Price et al., 1988	50	Dominant	.4%	Males	.789	.789	.004	Males	54%
				Females	.763	.763	.003	Females	9%

*TT, Tt, and tt represent the homozygous affected, heterozygous affected, and homozygous unaffected genotypes, respectively.
†The single, major locus genetic model is abbreviated SML.

be more apt to appear with chronic tics, whereas genetically vulnerable females may more often appear with obsessive-compulsive symptoms).

Searching for the TS Gene Results from segregation analyses have indicated the plausibility of the involvement of a major gene in the etiology of TS. Currently, several groups of researchers in the United States as well as in Great Britain and The Netherlands are conducting linkage analyses of multigenerational TS families.

Although TS has not yet been consistently linked to any specific region of the genome, several areas have been intriguing. For example, in 1987, Comings et al. reported a family in which six affected members all carried a t(7;18)(q22:q22) balanced translocation. This was followed by the report of a woman suffering from obsessive-compulsive behavior and other severe behavioral problems who had a deletion at 18q22.2. Linkage analyses in a British pedigree, however, excluded the association of this region on chromosome 18 with the transmission of TS (lod score = −2.74). Similarly, linkage between a region on the short arm of chromosome 3 and TS was tested and excluded after the observation of a t(3:8)(p21.3;q24.3) translocation in an affected individual.

Using a mini-satellite probe, researchers at Washington University in St. Louis have detected a fragment that appears to be segregating with five affected males and three obligate carrier females in one four-generation family. The fragment has been isolated from one of the affected individuals and will be used in further studies as a linkage probe for TS in this family.

In addition to the analyses discussed above, several investigators have been conducting linkage analyses utilizing known genetic markers scattered throughout the genome. To date, 228 genetic markers have been screened in a series of 10 affected multigenerational families from the United States, Canada, The Netherlands, and Norway, and more than 50% of the genome has been excluded thus far. A recent report claimed possible linkage of TS to genetic markers on the short arm of chromosome 3. Several methodological difficulties inherent in the analyses and the reporting of results, however, make this claim premature.

PATHOPHYSIOLOGY

Neurophysiology

The pathophysiology of TS remains unknown despite the wealth of information that has been discerned from pharmacological data and peripheral neurochemical data. In other words, there is as yet no etiological basis that dictates treatment.

Although no comprehensive etiological theory exists as yet, there is considerable evidence to suggest a central role of dopaminergic mechanisms in the pathophysiology of TS. The first line of evidence comes from the repeated observation of lowered cerebral spinal fluid levels of homovanillic acid, a major metabolite of brain dopamine, in TS patients as compared with controls. The second line of evidence comes from empirical drug trials in patients with Tourette's syndrome. For example, dopaminergic-blocking agents, such as haloperidol, have been found to be somewhat effective for controlling tics. Similarly, inhibitors of dopamine synthesis have been found to be effective in controlling the symptoms of TS in some patients. Conversely, tics are often exacerbated after administration of agents that increase dopaminergic activity, and TS-like symptoms have been observed in a small number of individuals following withdrawal of neuroleptics.

Although the preceding evidence is intriguing, it does not fully explain the pathophysiology of TS. Whereas many TS individuals benefit from dopaminergic agents, not everyone does. Furthermore, these drugs result most often in tic suppression but not elimination. Although etiological heterogeneity may exist, it does not appear, for even a subset of TS individuals, that central dopaminergic mechanisms are the sole factor in the expression of TS.

Several other neurochemical mechanisms have been implicated in the development of TS. Limited success from drug efficacy trials with clonidine, clonazepam, and cholinergic agents has led to speculation concerning the roles of the noradrenergic, GABAergic, and cholinergic systems. Peripheral neurochemical data weakly implicate serotonergic mechanisms. This is particularly interesting given the potential role of serotonergic mechanisms in obsessive-compulsive disorder and the suggested relationship, previously discussed, between OCD and TS. Medications that alter serotonergic activity, however, do not produce consistent effects on TS symptoms.

Neuroanatomy

Results from the aforementioned neurochemical studies are indicative of central nervous system dysfunction in TS. Neuroanatomical studies, however, have not been able to discover any anatomical correlates of the clinical manifestations of this disorder.

Abnormal EEGs, generally increased slow wave and posterior spikes, have been reported in approximately 50% of TS patients studied. Earlier reports suggested the rate of abnormal computerized brain tomography scans was approximately 25% in TS patients, with an association between the likelihood of abnormalities on the scan and the EEG. More recently, however, only two abnormal scans were observed from a total of 90 TS patients. Studies of main structure and function utilizing imaging techniques such as positron emission tomography, magnetic resonance imaging, and single-photon emission tomography are now being performed in TS populations.

TREATMENT

Effective management of the TS individual depends on several factors, including the severity of the symptoms and the patient's perception of these symptoms, as well as the recognition and support the individual receives from family and community. Assessing the efficacy of treatment is inherently problematic, however, due to the very nature of tic behavior. Specifically, difficulties are encountered in trying to determine how to measure behavior after treatment and in what setting(s), so that follow-up yields meaningful results. The treatments most generally used for TS fall into three categories: psychotherapy, behavioral therapy, and pharmacological therapy.

Psychotherapy

Psychotherapy has not proven to be a particularly effective method for treating TS. For example, in a recent study of 75 TS individuals, 73% reported seeking treatment from a psychiatrist or psychologist. Of these, 27% had undergone psychoanalysis, with the majority of these individuals (75%) reporting that such treatment had no effect, either positive or negative.

Behavioral Therapy

Several types of behavioral therapy have provided limited benefits to TS patients. The technique most frequently employed, massed practice, is based upon the notion that tics are a conditioned response to stress that is then reinforced by a reduction in anxiety after tic expression. Massed practice of tic behavior is encouraged in order to build up conditioned inhibition and, subsequently, to extinguish the response. Other behavioral therapies include contingent positive reinforcement for tic-free periods, occasionally in combination with time-out or aversive consequences, interventions based upon habit reversal or symptom substitution, and self-monitoring and relaxation techniques.

Pharmacological Therapy

As indicated in a previous section, many drugs have been used in the treatment of TS. The most widely used drug is the dopamine-blocking agent haloperidol. This neuroleptic seems to have a more specific effect than merely anxiety reduction, as other neuroleptics, such as tranquilizers and sedatives, provide only temporary relief from TS symptoms.

Haloperidol is beneficial to the majority of TS patients. Recently,

Jagger et al. reported that 91% of her sample had taken haloperidol at some time. Eighty-eight percent of these individuals had experienced a reduction in symptoms. However, 80% also experienced one or more undesirable side effects. Dystonic side effects are often experienced but can generally be managed by the additional prescription of an anti-parkinsonian agent. Other side effects of haloperidol include depression, headaches, weight gain, and motivational and cognitive dulling.

Because of the numbers of patients experiencing unpleasant side effects from haloperidol, the efficacy of other agents in the treatment of TS has been explored. Another more selective dopamine-blocking agent, pimozide, has been found to be advantageous for tic control, although it appears to be slightly less effective than haloperidol. Patients reportedly experience less sedation with the use of pimozide as compared with haloperidol; however, depression is still a pronounced side effect. There is also some concern about the risk of electrocardiogram abnormalities with the administration of pimozide.

The efficacy of the noradrenergic agent clonidine is somewhat unclear, but there are indications that it may be beneficial for some TS patients. Fairly long-term administration of the drug (12 to 52 weeks) is necessary in order to observe symptom improvements. Side effects include sedation, withdrawal effects, and an apparent risk of tolerance with long-term use.

Several other drugs have been used with limited success in the treatment of TS, including chlorimipramine, desipramine, physostigmine, tetrabenazine, clonazepam, fluphenazine, lithium, and trazodone. In general, drug efficacy studies are difficult to interpret due to the natural waxing and waning course of TS symptoms, as well as placebo effects. For these reasons, randomized, double-blind studies with adequate controls and sufficiently large sample sizes are needed to evaluate properly the efficacy of various pharmacological treatments for TS.

SUMMARY

Tourette's syndrome is a serious disorder involving the interaction of neurophysiological, genetic, behavioral, and environmental factors. The clinical picture of Tourette's syndrome has been well characterized since its initial description in 1885; however, only recently have significant advances been made with regard to understanding the etiology of the disorder. Much remains to be learned about the disorder, including the clinical range of phenotypic expression, its neurophysiology, and its etiology.

Because the symptomatology of TS is varied, both among individuals and within individuals across time, the range of expression for TS is currently an area of considerable interest. For example, obsessive-compulsive disorder is found much more frequently in TS individuals than in the general population. This association, however, is not sufficient to establish if TS and OCD are etiologically related. It has been found, however, that the rate of OCD is equal in relatives of TS index cases regardless of whether or not the index case also received a diagnosis of OCD. Therefore, TS and OCD may represent alternative expressions of the same disorder within these families. Because there have been a variety of other disorders associated with TS, the question remains as to what other behaviors constitute the true range of expression of TS. Proper definition of the disorder will aid etiological studies as well as act as a guide for treatment.

Future investigations regarding TS also have the exciting possibility of discovering a major gene responsible for the transmission of the disorder. Once this has been accomplished, attempts will be made to isolate the gene and identify the gene product. An understanding of the genetics of the disorder will, in the short-term, help establish more accurate risk estimates for relatives of TS individuals and, in the long-term, aid the development of effective treatment strategies. In addition, such knowledge will begin to provide a basis for a better understanding of environmental factors important in the development of TS. From previously conducted segregation analyses, it appears that individuals at equal risk for developing TS do not always develop the disorder or express the disorder to the same degree. Once linked genetic markers are available, carefully designed at-risk studies may begin to identify environmental variables crucial to the development of TS.

TS is a prime example of a neuropsychiatric disorder that has benefitted from a multidisciplinary research approach. The combined information regarding clinical description, epidemiology, pathophysiology, and genetics has led to a better understanding of the TS individual. Future research promises to continue to be interdisciplinary in nature, drawing basic and clinical researchers to the intriguing question of the relationship between the biology of the mind and behavior.

ACKNOWLEDGMENTS

This work was supported in part by grants NS-16648, MH-00508 (a Research Scientist Development Award to Dr. Pauls), and MH-18268 (a postdoctoral training grant supporting Dr. LaBuda).

SELECTED REFERENCES

American Psychiatric Association. Diagnostic and statistical manual of mental disorders. 3rd ed. Washington, D.C.: American Psychiatric Association, 1980.

Azrin NH, Peterson AL. Behavior therapy for Tourette's syndrome and tic disorders. In: Cohen DJ, Bruun RD, Leckman JF, eds. Tourette's syndrome and tic disorders: clinical understanding and treatment. New York: John Wiley & Sons, 1988.

Baron M, Shapiro E, Shapiro A, Rainer JD. Genetic analysis of Tourette syndrome suggesting major gene effect. Am J Hum Genet 1981;33:767–775.

Brett P, Curtis D, Gourdie A, et al. Possible linkage of Tourette syndrome to markers on short arm of chromosome 3 (C3p21-14). Lancet 1991;338:1076.

Bruun RD. The natural history of Tourette's syndrome. In: Cohen DJ, Bruun RD, Leckman JF, eds. Tourette's syndrome and tic disorders: clinical understanding and treatment. New York: John Wiley & Sons, 1988.

Cohen DJ, Detlor J, Shaywitz B, Leckman JF. Interaction of biological and psychological factors in the natural history of Tourette syndrome: a paradigm for childhood neuropsychiatric disorders. In: Friedhoff AJ, Chase TN, eds. Advances in neurology: Gilles de la Tourette syndrome. New York: Raven Press, 1982.

Comings DE, Comings BG, Devor EJ, Cloninger CR. Detection of major gene for Gilles de la Tourette syndrome. Am J Hum Genet 1984;36:586–600.

Comings DE, Comings BG, Dietz G, et al. Evidence that the Tourette syndrome gene is at 18q22.1. Seventh International Congress on Human Genetics, Berlin, 1987;II:620.

Devor EJ. Complex segregation analysis of Gilles de la Tourette syndrome: further evidence for a major locus mode of transmission. Am J Hum Genet 1984;36:704–709.

Devor EJ, Burgess AK, Isenberg KE. Random mapping of Tourette syndrome with a hypervariable "minisatellite" probe: identification of a cosegregating restriction fragment. Am J Hum Genet 1989;45:A183.

Donnai D. Gene location in Tourette syndrome. Lancet 1987;1(8533):627.

Fahn S, Erenberg G. Differential diagnosis of tic phenomena: a neurologic perspective. In: Cohen DJ, Bruun RD, Leckman JF, eds. Tourette's syndrome and tic disorders: clinical understanding and treatment. New York: John Wiley & Sons, 1988.

Gilles de la Tourette G. Etude sur une affection nerveuse characterisee par de l'incoordination motrice accompagnee de echolalie et de copralalie. Arch Neurol 1885;8:68–84.

Goetz CG, Klawans HL. Gilles de la Tourette on Tourette syndrome. In: Friedhoff AJ, Chase TN, eds. Advances in neurology: Gilles de la Tourette syndrome. New York: Raven Press, 1982.

Heutink P, Sandkuyl LA, van de Wetering BJM, et al. Linkage and Tourette syndrome. Lancet 1991;337:122–123.

Jagger J, Prusoff BA, Cohen DJ, Kidd KK, Carbonari CM, John K. The epidemiology of Tourette's syndrome: a pilot study. Schiz Bull 1982;8:267–278.

Kidd KK, Pauls DL. Genetic hypotheses for Tourette syndrome. In: Friedhoff AJ, Chase TN, eds. Advances in neurology: Gilles de la Tourette syndrome. New York: Raven Press, 1982.

Leckman, JF, Cohen DJ. Descriptive and diagnostic classification of tic disorders. In: Cohen DJ, Bruun RD, Leckman JF, eds. Tourette's syndrome and tic disorders: clinical understanding and treatment. New York: John Wiley & Sons, 1988.

Leckman JF, Price RA, Walkup JT, Ort S, Pauls DL, Cohen DJ. Nongenetic factors in Gilles de la Tourette's syndrome. Arch Gen Psychiatry 1987;44:100.

Leckman JF, Riddle MA, Cohen DJ. Pathobiology of Tourette's syndrome. In: Cohen DJ, Bruun RD, Leckman JF, eds. Tourette's syndrome and tic disorders: clinical understanding and treatment. New York: John Wiley & Sons, 1988.

Leckman JF, Walkup JT, Cohen DJ. Clonidine treatment of Tourette's syndrome. In: Cohen DJ, Bruun RD, Leckman JF, eds. Tourette's syndrome and tic disorders: clinical understanding and treatment. New York: John Wiley & Sons, 1988.

Moldofsky H, Sandor P. Pimozide in the treatment of Tourette's syndrome. In: Cohen DJ, Bruun RD, Leckman JF, eds. Tourette's syndrome and tic disorders: clinical understanding and treatment. New York: John Wiley & Sons, 1988.

Pakstis AJ, Heutink P, Pauls DL, et al. Progress in the search for genetic linkage with Tourette syndrome: an exclusion map covering more than 50% of the autosomal genome. Am J Hum Genet 1991;48:281–294.

Pasmanick B, Kawi A. A study of the association of prenatal and paranatal factors in the development of tics in children. Pediatrics 1956;48:596–601.

Pauls DL, Cohen DJ, Heimbuch R, Detlor J, Kidd KK. Familial pattern and transmission of Gilles de la Tourette syndrome and multiple tics. Arch Gen Psychiatry 1981;38:1091–1093.

Pauls DL, Hurst CR, Kruger SD, Leckman JF, Kidd KK, Cohen DJ. Gilles de la Tourette's syndrome and attention deficit disorder with hyperactivity: evidence against a genetic relationship. Arch Gen Psychiatry 1986;43:1177–1179.

Pauls DL, Kruger SD, Leckman JF, Cohen DJ, Kidd KK. The risk of Tourette's syndrome and chronic multiple tics among relatives of Tourette's syndrome patients obtained by direct interview. J Am Acad Child Psychiatry 1984;23:134–137.

Pauls DL, Leckman JF. The inheritance of Gilles de la Tourette's syndrome and associated behaviors: evidence for autosomal dominant transmission. N Engl J Med 1986;315:993–997.

Pauls DL, Leckman JF. The genetics of Tourette's syndrome. In: Cohen DJ, Bruun RD, Leckman JF, eds. Tourette's syndrome and tic disorders: clinical understanding and treatment. New York: John Wiley & Sons, 1988.

Pauls DL, Towbin KE, Leckman JF, Zahner GEP, Cohen DJ. Gilles de la Tourette's syndrome and obsessive-compulsive disorder: evidence supporting a genetic relationship. Arch Gen Psychiatry 1986;43:1180–1182.

Price RA, Kidd KK, Cohen DJ, Pauls DL, Leckman JF. A twin study of Tourette's syndrome. Arch Gen Psychiatry 1985;42:815–820.

Price RA, Pauls DL, Kruger SD, Caine ED. Family data support a dominant major gene for Tourette syndrome. Psychiatric Res 1988;24:251–261.

Riddle MA, Hardin MT, Ort SI, Leckman JF, Cohen DJ. Behavioral symptoms of Tourette's syndrome. In: Cohen DJ, Bruun RD, Leckman JF, eds. Tourette's syndrome and tic disorders: clinical understanding and treatment. New York: John Wiley & Sons, 1988.

Robertson MM, Trimble MR, Lees AJ. The psychopathology of the Gilles de la Tourette syndrome. Br J Psychiatry 1988;152:383–390.

Shapiro A, Shapiro E. Tourette syndrome: history and present status. In: Friedhoff AJ, Chase TN, eds. Advances in neurology: Gilles de la Tourette syndrome. New York: Raven Press, 1982.

Shapiro AK, Shapiro E. Treatment of tic disorders with haloperidol. In: Cohen DJ, Bruun RD, Leckman JF, eds. Tourette's syndrome and tic disorders: clinical understanding and treatment. New York: John Wiley & Sons, 1988.

Shapiro E, Shapiro AK, Fulop G, et al. Controlled study of haloperidol, pimozide, and placebo for the treatment of Gilles de la Tourette's syndrome. Arch Gen Psychiatry 1989;46:722–730.

Stokes A, Bawden HN, Camfield PR, Backman JE, Dooley JM. Peer problems in Tourette's disorder. Pediatrics 1991;87:936–942.

TSA Permanent Research Fund. Fourth genetic workshop on Tourette syndrome. Bayside, New York: Tourette Syndrome Association, 1989.

Zahner GEP, Clubb MM, Leckman JF, Pauls DL. The epidemiology of Tourette's syndrome. In: Cohen DJ, Bruun RD, Leckman JF, eds. Tourette's syndrome and tic disorders: clinical understanding and treatment. New York: John Wiley & Sons, 1988.

CHAPTER 10

The Tuberous Sclerosis Complex

M. R. Gomez
Moyra Smith

Tuberous sclerosis complex (TSC) is a disorder of cell migration, proliferation, and differentiation involving practically all tissues of the human body. Notable exceptions are the spinal cord, which is rarely if ever affected, the peripheral nerves, muscles, pituitary, and pineal gland, which has never been reported to be involved.

HISTORY

In 1835 under the title "vascular vegetations," Rayer's atlas of dermatological diseases depicted a patient whose facial lesions we can recognize today as typical facial angiofibromas. There is no additional information given with this colored drawing, most likely the first published case of the tuberous sclerosis complex. Von Recklinghausen, in 1864, presented to the Obstetrical Society of Berlin a newborn who had died after taking only a few breaths. At postmortem there were several cardiac "myomata" embedded in the myocardium or protruding into the cavities, and "several scleroses" of the brain. In 1880, D-M. Bourneville described the cerebral lesions he found at postmortem in a 15-year-old girl who had died in status epilepticus, for which he coined the term "tuberous sclerosis of the cerebral circumvolutions." Tuberous sclerosis or TS is the term used today not only for the cerebral pathology of the disease but also for the disease itself, manifested by the presence in multiple organs of a variety of hamartomas and hamartias. Tuberous sclerosis complex or TSC is a better name for the entire clinical and pathologic constellation of this protean disease, which is first clinically recognized by its skin lesions and particularly by facial angiofibroma (improperly called adenoma sebaceum). The association with renal tumors was first

215

reported by Bourneville in 1880 and again by Bourneville and Brissaud in 1881.

From early in this century until two decades ago, the association of facial angiofibroma with seizures and mental retardation, also known as the Vogt triad, was the only clinical indicator of cerebral tuberous sclerosis. The frequent observation that retinal hamartomas could be found in these patients added a new diagnostic feature of great value. With the successive introduction of simple radiography, pneumoencephalography, ultrasound (US), computed tomography (CT), echocardiography, and magnetic resonance imaging (MRI) as diagnostic techniques, cerebral, cardiac, and renal lesions of TSC have been recognized in patients and asymptomatic relatives, and the prevalence of TSC has been found to be 10 times greater than originally estimated. It is possible that currently we do not identify certain individuals with TSC because their lesions are not revealed with our present examination methods. The diagnostic criteria have gradually changed with the discovery of signs obtained by imaging the brain and other organs.

PROPOSED DIAGNOSTIC CRITERIA

We propose that the clinical and imaging features selected for diagnosing TSC be segregated according to their practical value into one of the following three categories: definitive, presumptive, or suggestive diagnostic features, as shown in Table 10.1.

It is impossible at this stage of our knowledge to completely exclude the diagnosis of TSC from any person, particularly from those at risk because of an affected direct relative. The absence of all the features listed in Table 10.1 in an individual at risk who has been properly examined makes it extremely unlikely that he or she is affected. In practice, the organs that need to be examined directly or by imaging are the skin, retina, brain, and kidneys. Less commonly, the bones are radiographically examined and the teeth are examined after staining their surface to detect enamel pits. It is not always practical to examine all organs and it is impossible to detect the smaller hamartomas in the organs examined.

For the purpose of genetic counseling it is accepted that subjects without any of the features listed in Table 10.1 have less than one percent (1%) probability of being affected even if they have a direct relative with TSC. Skipped (or bracketed) generations are a rarity in TSC; only a few families have been reported in which neither parent of two or more affected children had signs of TSC.

Table 10.1:   Tuberous Sclerosis, Hierarchy of Diagnostic Features

Organ	Definitive	Presumptive	Suspect
CNS	*Cortical tubers* *Subependymal nodules* *Giant-cell astrocytoma*		Infantile spasms Generalized/ partial seizures
Retina	Hamartomas	Hamartoma	Iris/eyelashes depigment
Skin	Facial angiofibromas Ungual fibroma Fibrous forehead plaque	Confetti-like spots	Hypomelanotic macules
Kidneys	*Multiple angiomyolipomas*	Angiomyolipoma	Cysts
Heart		Multiple rhabdomyomas	Rhabdomyoma
Lungs		Lymphangio- myomatosis	Spontaneous pneumothorax Chylothorax Honeycomb image
Teeth			Enamel pits
Gingiva			Fibromas
Rectum		Polyps	
Thyroid			Adenoma
Adrenal			Angiomyolipoma
Gonads			Angiomyolipoma
Liver			Angiomyolipoma
Bones			Cysts

Listed vertically under each category are the majority of, but not all, clinical features of TSC. Horizontally to the right of the organ name are the lesions found in that organ. A feature's name printed in italics indicate these are lesions that sometimes may be in question; when only a CT, MR, or US image is available, this image should be unequivocally identified. The indirect viewing of a lesion as an image on a radiograph, US, CT scan, or MRI may not be distinct enough to establish the diagnosis. In this situation, two of these features listed in italics would be necessary to establish the diagnosis of TSC. This is to say that a single feature of those listed under the heading "Definitive" is sufficient for the diagnosis of TSC only if the image in question truly corresponds to the lesion technologically recognized as one of TSC.

For a presumptive diagnosis of TSC, only one of the features listed under "Presumptive" is necessary. If an individual has two of the features listed under the heading "Presumptive" or only one of these features plus a direct relative with an established definitive diagnosis, the diagnosis of TSC is definitive. On the other hand, with one or more of the features listed under the heading "Suggestive," the diagnosis of TSC can only be suspected. It is prudent that subjects with one of the features listed under the heading "Suspect" should be classified as "Presumptive" when they have a direct relative with the diagnosis of definitive TSC.

CENTRAL NERVOUS SYSTEM INVOLVEMENT

The pathologic expression of the defective gene or genes of TSC is primarily in the form of well-circumscribed lesions within the involved organs. It is a unique characteristic of TSC that such pathology does not entirely involve the affected organ but is confined to only one or more regions in that organ, usually in the form of hamartias and hamartomas and rarely as hamartoblastoma. Small lesions usually cause no symptoms. Lesions may replace enough normal tissue to cause organ failure or dysfunction when numerous and large. Contrary to what occurs in malignant neoplastic processes, the hamartomas do not infiltrate or metastasize, although they may displace and compress normal adjacent tissue as they grow.

The great variation of clinical features among patients undoubtedly depends on which organ(s) is/are involved. Affected individuals from different families show greater discrepancy among them than individuals from the same family.

Symptoms

Seizures The most frequent clinical manifestation of patients with TSC is epileptic seizures, alone or in association with mental subnormality, abnormal behavior, or autistic features. Mental subnormality does not occur in patients with TSC who have never had seizures, but not all patients with TSC and seizures are mentally subnormal. The seizures may be partial or of any generalized type except for typical absences. Neonates with TSC most frequently present with partial motor, myoclonic, or tonic seizures, while infants older than two months present with infantile spasms or myoclonic seizures. After the first year of life complex partial seizures originating from frontal or temporal regions with or without generalization are common. Seizures starting after the first year of life may be complex partial, atonic, tonic, or myoclonic.

Arrest or regression of psychomotor development is not unusual in patients whose seizures started within the first five years of life and were frequent. A direct correlation exists between the presence of large and numerous cortical tubers and an early onset of seizures that increase in frequency and severity. These patients as a rule fail to attain normal mental development.

Intracranial Hypertension A different clinical form of presentation is with symptoms and signs of increased intracranial pressure. This is almost always due to a subependymal giant-cell astrocytoma (SEGA) growing into a lateral ventricle near the foramina of Monro and blocking one or both, thus causing dilatation of one or both lateral ventricles. A SEGA growing into the ventricle may form an intraventricular cast and cause only partial obstruction for years before symptoms of increased intracranial pressure are recognized. With a more rapid and complete blockage there is an abrupt onset of symptoms.

Examination of the patient's eyegrounds may disclose chronic or acute papilledema. A head CT scan or MRI will reveal dilated ventricles. Generally, obstruction of the cerebrospinal fluid (CSF) circulation by a SEGA occurs between the ages of 5 and 22 years.

Cerebellar ataxia, hemiplegia, hemianopsia, choreoathetosis, autism, and progressive dementia are less common clinical features than intracranial hypertension. Autism and progressive dementia only occur when patients with TSC have had seizures.

Neuropathology

Four types of cerebral lesions may be found in this disorder: cerebral tubers or tuberosities, subependymal nodules, giant-cell tumors or astrocytomas (SEGA), and nests of heterotopic neurons within the white matter.

The cortical tuber is a cerebral hamartia, that is, dysplastic cerebral cortex and subjacent white matter. Tubers are multiple and visible on gross inspection as widened gyri protruding slightly over the cerebral surface. They are slightly pale and firmer to palpation than the surrounding normal brain. Their number, size, and location vary a great deal from one patient to another. When small, they are easier to locate visually than by palpation. Some tubers are large enough to involve two or more adjacent gyri and intervening sulci. They are less common in the cerebellum than in the brain.

On microscopic examination the tubers display a striking disorganization of the ganglion cell arrangement and loss of the normal laminal pattern. The neuronal nuclei are reduced and the astrocytic nuclei are increased in number. Some neurons are disoriented in relation to the cortical surface and positioned horizontally or even vertically, with the apical dendrite pointing away from the pial surface. Among the most prominent changes in pyramidal cells are shrinkage, chromatolysis, glycogen accumulation, and an excess of lipopigment. There are also large cells whose characteristics are neither clearly neuronal nor astrocytic and whose origin has been the subject of much controversy. Immunohistochemical studies have shown clusters of the large cells stained for glial fibrillary acidic protein (GFAP), a marker for astrocytes, next to clusters of other large cells that do not take the GFAP stain. With the Golgi-Cox method, it has been shown that the neurons of cortical tubers have an abnormal morphology, with unusual dendrites, that are often spineless or have a reduced number of spines and a beaded appearance or varicosities. The latter finding, although prominent in fetal neurons, is uncommon in the postnatal cerebral cortex.

Subependymal nodules are seen on the ventricular walls as rounded protrusions into the ventricular cavity that resemble candle gutterings. They are most often found along the course of the stria terminalis and near the foramina of Monro. They also occur, though less often, in the third and fourth ventricles and in the Sylvian aqueduct. These nodules are formed by abnormally appearing astrocytes of fusiform or plump appearance covered by intact ependyma. These astrocytes vary in size and may be extremely large and multinucleated. The subependymal nodules often have a prominent vascular stroma with thick vessel walls. In this gliovascular stroma there may be concentric microspherules or calcospherites. Hemorrhage and necrosis are uncommon findings.

The third pathological element found in the brain of TSC patients, the giant-cell astrocytoma (GCA), is a hamartoma with the same histological characteristics and often the same location as the subependymal nodule. The only difference between a nodule and GCA is continued growth of the latter, causing symptoms of CSF obstruction as aforementioned. Some GCAs are found within the white matter, presumably originating from clusters of undifferentiated heterotopic cells located along the migratory path of primordial glial cells and neurons. The GCAs contain few mitotic figures and are not malignant, although their appearance may suggest otherwise.

Neuroimaging

The calcified subependymal nodules and GCAs can be displayed on plain radiographs. Pneumoencephalography and ventriculography were used in the past to detect hamartomas within the ventricles. The image called "candle guttering" seen in the pneumoencephalograms became the most revealing sign when searching for cerebral hamartomas. The newer noninvasive imaging methods, CT and MRI, have replaced air encephalography.

CT scanning may demonstrate images of any of the three types of cerebral lesions: the subependymal nodules, GCAs, or cortical tubers. The subependymal nodules are best seen in noncontrasted head CT scans after they have become calcified, usually not before the patient is five months old. The calcifications are in the subependymal region on the lateral ventricles, protruding into them or imbedded into the caudate nucleus or thalamus. They are asymmetrical. Their location and type of image are so characteristic that it is easy to differentiate them from venous angiomas, cysticercosis, and prenatal inflammatory lesions caused by cytomegalovirus or toxoplasma capsulatum. Heterotopic gray matter along the external wall of the ventricles may simulate uncalcified subependymal nodules.

The MRI on T-1 weighted sequences reveals the subependymal nodules as small projections into the ventricles isointense to white matter and slightly hyperintense to gray matter. On the T-2 weighted scans the nodules are isointense or hypointense to the white and gray matter and contrast well with the hyperintense CSF. After calcification the subependymal nodules give a less intense signal on both T-2 and T-1 weighted images, thus facilitating their recognition.

Subependymal giant-cell astrocytomas arise from subependymal nodules located on the inferior part of the head of the caudate nucleus, and may grow into the ventricular cavity to occlude one or both foramina of Monro. In CT scanning, these tumors enhance with intravenous injection of contrast media, thus appearing as bright white masses due to an intensely increased attenuation of the x-rays. In MR scanning, gadolinium injection enhances these tumors.

Cortical tubers may be detected on the uncontrasted CT scan by a focal decreased attenuation of the subcortical white matter. The cortex itself may appear as an area of increased attenuation if there has been some degree of calcification. In serial CT scans obtained through the years, the white matter underlying the cortical tubers will display an increase in attenuation and become less distinct from the surrounding normal white matter. Subcortical areas of decreased attenuation in the tubers are hypomyelinated histopathologically.

In the first months of the patient's life, when the normal white matter of brain is still insufficiently myelinated, the hypomyelinated subcortical white matter of the tuberous gyri does not stand out well with CT or MR scanning. As the patient gets older, and certainly by the end of the first year, some cortical tubers are detectable by their hypomyelinated area. With MRI in T-1 sequences, cortical tubers have the appearance of "empty gyri": a dark central core surrounded by an isointense ring. In T-2 sequences, there is hyperintensity of the tuber's subcortical region. In large tubers, the hyperintense area is extensive and may connect one gyrus with the next by involving the subcortical white matter of the intervening sulcus.

Radial low-attenuation bands or streaks extending from the ventricular region to the cortex, seen with the CT scan and better identified with MRI, indicate a migration disorder along these lines where clusters of heterotopic undifferentiated cells remain and myelination is impaired. Within these radial bands may be found a growing GCA.

EXTRANEURAL INVOLVEMENT

Extraneural expression of the TSC gene(s) occurs chiefly in skin, kidneys, heart, large arteries, and lungs. The endocrine glands, teeth, gums, gastrointestinal tract, and bony skeleton may also be affected.

Skin   Skin lesions seen in TSC include facial angiofibromas, periungual fibromas, fibrous forehead plaques, shagreen patches, and hypomelanotic macules. The first three lesions are pathognomonic of this disease. When any of these three are found and are unequivocal, the diagnosis is unquestionable.

The facial angiofibromas usually appear around the age of 3 years and very rarely after puberty. Thus, they are present in more than 50% of adult patients only and in 30% of patients if children are included. Two other skin hamartomas, the ungual fibroma and the shagreen patch, appear in the second decade of life and are found in approximately 20–30% of TSC patients of all ages. Although the ungual fibroma is pathognomonic of TSC, the shagreen patch may not be so. The shagreen patch consists of a cluster of connective tissue hamartomas histologically similar to facial angiofibromas and to ungual fibromas; it may need histologic confirmation.

The fibrous plaque, another hamartoma most often found on the forehead, scalp, eyelids, or cheeks, is also histologically similar to the facial angiofibroma. Since it sometimes is seen in the newborn period, it is the earliest clinical sign to be found that is pathognomonic for TSC. These plaques may grow through the years and sometimes calcify.

The white spot, or hypomelanotic macule (HM), is the most prevalent of all the skin findings in TSC patients. They may be lance-ovate, round, or irregular in shape and vary in diameter between 2 mm and several centimeters. About 90% of subjects with TSC will have more than four HMs. The presence of one or two HMs alone is not sufficient to make the diagnosis of TSC on an individual, even if he/she is at risk by having a direct relative with a definitive diagnosis of TSC. Although not every subject with TSC has white spots and not every person with white spots has TSC, the spots are very convenient for making a presumptive diagnosis. A group of small confetti-like white macules has been reported to be sufficient for making a provisional diagnosis of TSC unless they appeared after prolonged and repeated sun exposure of the skin.

The histopathology of all four hamartomas is strikingly similar: dermal fibrosis with sclerosis and layering of the collagen fibers haphazardly arranged. Only the facial angiofibroma and fibrous forehead plaques have increased vascularity and dilated vessels, giving the lesions a reddish discoloration and fleshy appearance. In these hamartomas, the dermis contains spider-shaped cells with the appearance of glial cells. These cells proliferate in tissue culture and maintain the same stellate configuration and glial appearance.

Histopathologically, the hypomelanotic macule is characterized by the presence of a normal number of melanocytes. The melanosomes are reduced in size and the melanin is reduced or absent as determined by electromicroscopic examination and histochemical reaction for dopa of melanocytes and keratinocytes.

Other skin lesions found in TSC patients are of lesser importance. Café-au-lait spots are seen with greater frequency in TSC patients than in the normal population, as are skin tags or molluscum fibrosum pendulum (soft pedunculated fibromas found on the neck of patients with TSC).

Kidneys   Renal involvement is second to neural involvement as a cause of morbidity and mortality in TSC patients. There are three types of renal lesions in TSC: angiomyolipomas (AML), renal cysts, and renal cell carcinomas.

Angiomyolipoma, a renal hamartoma, is found in 80% of TSC patients who come to autopsy and has been detected in 45% of living individuals with TSC whose kidneys were examined with imaging methods, the majority of whom were asymptomatic.

Multiple renal angiomyolipomas are pathognomonic of TSC. However, a single renal angiomyolipoma is not sufficient for the diagnosis unless it is associated with multiple renal cysts.

Angiomyolipomas, whether associated or not with TSC, have a characteristic appearance. They often bulge from the renal surface as yellow rounded masses and are seen in cut sections of the kidney as yellowish solid tumors. Although they do not infiltrate the normal parenchyma, they displace it and may penetrate the renal capsule and extend into the perirenal tissues or into the renal vein. Microscopically, they are made up of fat, smooth muscle, and blood vessels. In addition to fat cells with compressed nuclei, there are scattered foamy polygonal cells with central nuclei, sometimes multinucleated, and prominent nucleoli. The smooth muscle cells are in disorganized sheets or clusters between fat cells or form concentric layers around the blood vessels. The blood vessels within the tumor are large, thick-walled, and resemble arteries except for the lack of an elastic layer.

Renal symptoms or signs of AML rarely appear before the third decade of life. Symptoms include flank pain, hematuria, hypertension, and uremia. The most dreaded event is sudden bleeding into the kidney from ruptured aneurysmatic vessels within the AML followed by bleeding into the retroperitoneal space. Prompt treatment of hypovolemic shock and nephrectomy may save these patients. Persistent AML bleed in the form of hematuria is treatable with arterial embolization.

The second most common renal finding in TSC is multiple cysts. These are also asymptomatic except when large and numerous. Renal cysts are found in at least 30% of patients with TSC. In rare exceptions, the cysts are so numerous and the displacement of renal parenchyma is so extensive that renal failure results. The few patients thus affected have been young children who developed uremia and hypertension. When renal cysts are combined with even a single renal angiomyolipoma, it is unequivocal that the patient has TSC.

The cysts are lined with hyperplastic epithelium with large acidophilic cells containing hyperchromatic nuclei and occasional mitotic figures. Hyperplastic cells from this epithelium form small intratubular masses that are characteristic of TSC and not found in patients with autosomal dominant polycystic disease of the kidneys.

Renal clear-cell carcinomas have been reported infrequently in patients with TSC. It is believed that these are hamartoblastomas originating in the hyperplastic epithelium of the renal cysts. The clear-cell carcinomas may metastasize.

Lungs Infrequent and almost exclusively confined to women in the third or fourth decade of life, pulmonary lymphangiomyomatosis (LAM) is more common in patients with TSC than in the general population. Trapping of alveolar air leads to pulmonary cyst formation, which may result in spontaneous rupture and pneumothorax. Blockage of lymphatics by hyperplastic perivascular smooth muscle fibers may cause chylothorax. Vascular rupture causes hemoptysis. The progression of pulmonary cyst formation with loss of pulmonary elasticity causes lung hyperinflation, respiratory failure, and hypercarbia. At this stage, the radiologic image of the lungs, with cysts surrounded by sclerosed interstitial walls, is reminiscent of a honeycomb. Progressive respiratory failure due to LAM is fatal in less than five years. Treatment consists of progesterone or lung transplantation.

The gross pathology of the lungs is very characteristic: they are large and twice as heavy as normal. On cut sections, the normal parenchymal pattern has been replaced by multiple cysts a few millimeters to several centimeters in diameter, giving the lung a spongiform appearance. The cysts are usually empty and their thin walls lack epithelial lining. The septi between the cysts contain immature-looking smooth muscle cells with ill-defined interdigitating lymphatic spaces. There is no fibrous tissue. These lesions of TSC are indistinguishable from pulmonary LAM of patients not known to have TSC.

Heart and Vascular System Cardiac involvement is in the form of cardiac rhabdomyoma, a type of hamartoma that is usually multiple and most often clinically silent. The rhabdomyomas are more often ventricular than atrial. They may be strictly intramural, protrude into the cardiac cavity, or bulge on the cardiac surface. On gross examination they have a gray or yellowish-white color, measure a few millimeters to several centimeters in diameter, and are well demarcated from the surrounding myocardium.

With the use of light microscopy the tumors appear well circumscribed, have no capsule, and are composed of glycogen-filled cells. When glycogen is washed out during fixation, the stained tissue will show a characteristic "chicken-wire" or "spider-cell" image. The cells are similar to normal cardiac Purkinje cells and may even function as such, forming an abnormal conducting bundle that causes the pre-excitation pattern characteristic of the Wolff-Parkinson-White syndrome.

In approximately 50% of infants with TSC rhabdomyomas are discovered by echocardiography. When symptomatic in neonates, they are manifested by signs of obstruction of flow at the ventricular outlet or the atrioventricular foramen. This obstruction may lead to cardiac failure in neonatal life. There have been a few cases of hydrops fetalis resulting in stillbirth or neonatal death from cardiac rhabdomyoma. Intramural hamartomas may also be associated with cardiac arrhythmia secondary to interruption of the specific conduction system. Heart failure may result from arrhythmia, from obstruction, or from replacement of contractile myocardial fibers by noncontractile tissue. The majority of rhabdomyomas of newborns or young infants with TSC tend to involute or decrease in size as the patient grows, as demonstrated in serial echocardiograms. Prenatal diagnosis of TSC has been possible with echocardiography by demonstrating fetal cardiac rhabdomyomas as early as the 26th gestational week.

Aneurysms of the descending thoracic or abdominal aorta, the subclavian, internal carotid, anterior cerebral, middle cerebral, or vertebral arteries occur with more frequency in children with TSC than is expected in this age group from congenital defects in the arterial wall. Subarachnoid hemorrhage, hemothorax from ruptured aortic aneurysm, and visual loss from giant aneurysm are rare manifestations of TSC.

Strokes attributed to embolization of cerebral arteries in infants harboring cardiac rhabdomyomas have been reported without either arteriographic or pathologic evidence. The hypothesis that emboli result from fragments detached from a tumor seems improbable; if embolization has indeed occurred it is more plausible that it was due to thromboemboli originating from clots formed in the blood turbulence created by the intracavitary rhabdomyomas.

Other Organs   Patients with TSC have dental enamel defects in the form of pits. Two or three pits per tooth surface is not unusual in these patients. These lesions are not pathognomonic of TSC and are only more frequently found in these patients than in the general population. The same may be said about gingival fibromas, a fibrous hamartoma. Macroglossia has been found infrequently. Fibromatous tumors may be found in the pharynx, larynx, and esophagus, and hamartomous polyps may be found in the rectal mucosa. Other hamartomas found in TSC patients are hemangioma of the spleen, angiomyolipoma of the adrenal glands, fibroadenoma of the testes, and adenoma of the thyroid. A variety of bone lesions occur in TSC. There may be cystic formation in the metatarsal and metacarpal bones; and there may be osteomatous thickening or sclerotic patches in the calvarium, vertebral bodies, pelvis, and long bones. These lesions are asymptomatic and their presence does not indicate that the subject necessarily has TSC.

Extraneural Imaging

Kidneys   The renal angiomyolipomas, due to their large amount of fat, produce a strong echo on ultrasonographic examination. This echodense property is not specific for angiomyolipomas since renal carcinomas have given the same image. Renal cysts are well demonstrated on ultrasound. The combination of angiomyolipomas and renal cysts in the same patient makes the diagnosis of TSC definitive.

CT examination of the abdomen demonstrates the fat in the angiomyolipomas. Renal cell carcinoma, a rare tumor in TSC, lacks fat. MRI also reveals fat with an intense signal on T-2 sequences and has the advantage of showing the vascular components of angiomyolipomas even without the administration of contrast material.

Heart   Two-dimensional echocardiography, MRI and, less often, angiocardiography are used for the detection of cardiac rhabdomyoma. Echocardiography can detect rhabdomyomas in the fetus after the 26th week of gestation.

Lungs   Plain radiographs will display the honeycomb appearance of lymphangiomyomatosis with cyst formation, and pneumothorax or chylothorax when they occur in patients with pulmonary TSC.

Bones   Cyst-like rarefaction of the metatarsal and metacarpal bones is a nonspecific finding in TSC, as is the osteomatous thickening or bone dysplasia found in the calvarium, vertebral bodies, pelvis, and long bones. The radiographic findings in the bone may be mistaken for Paget's disease and bone metastasis. If there is any question of diagnosis, the measurement of the serum alkaline phosphatase may be useful because it is normal in TSC and elevated in Paget's disease and bone metastasis.

GENETICS OF TUBEROUS SCLEROSIS

The population frequency of tuberous sclerosis was estimated as 1 in 10,000 by Wiederholt et al. in 1985. TSC is inherited as an autosomal dominant disease. A large proportion of cases, possibly more than 50%, are thought to represent new mutations since manifestations occur in children whose parents show no evidence of TSC. It is clear that at this time the incidence of new mutations in TSC and the population frequency of TSC cannot be accurately determined since there is a very high degree of variability of penetrance in TSC. An individual who carries the TSC gene may have no manifestations on clinical examination. In order to have a high degree of confidence that a person did not carry the TSC gene, it would be necessary to perform cranial CAT scan and MRI, renal ul-trasound or MRI, radiologic examination of the bones and lungs, and echocardiography, in addition to clinical examination. Despite the difficul-ties in assessing gene frequency and the incidence of new mutations due to nonpenetrance, the incidence of new mutations appears to be high.

Hypotheses to Explain the High Frequency of New Mutations

High frequency of new mutations could potentially be due to genetic heterogeneity, large gene size, or possibly, in certain sporadic cases of TSC, the interaction between two loci.

Hypotheses on Factors That Lead to Variable Penetrance in TSC

In order to understand the mechanism of penetrance versus nonpenetrance in TSC it will be necessary to determine the nature of the TSC gene mutation. One possible explanation for the TSC manifestations is that the TSC gene represents a tumor-suppressor gene and that an individual who has the TSC gene mutation does not develop a particular lesion unless a second mutation has taken place in a particular cell. The occurrence of extensive lesions in individuals who carry the TSC gene mutation may be due to a "second hit" occurring on the homologous chromosome in cer-tain progenitor cells early in development. Subsequent migration of the daughter cells of this progenitor cell to distant parts of a particular organ or to different parts of the body would lead to the occurrence of wide-spread lesions. The genetic event that constitutes such a second hit may be a structural chromosomal change or a mutation event, such as has been described in retinoblastoma. Although chromosomal deletion events have most frequently been associated with loss of tumor-suppressor genes, e.g., in retinoblastoma, it is clear that any structural chromosomal change that leads to disruption of the tumor-suppressor gene may result in a lesion.

If one proposes that the situation in TSC is analogous to that in retinoblastoma, the tumor-suppressor gene would act as a recessive gene. Sporadic cases of TSC would represent individuals in whom two hits had taken place in a particular cell or progenitor cell at some time during development. In the case of neurofibromatosis (NF), Xu et al. postulated that although a second mutational event may be required for development of neurofibromas, this second hit may not necessarily involve the NF gene but may involve another interactive gene, e.g., an oncogene.

There is evidence that suggests that other factors, e.g., hormonal factors, may affect expression of TSC manifestations. It is of interest to note that the lung lesions in TSC occur predominantly in women and are responsive to progesterone treatment. Certain TSC lesions are seldom seen before the onset of puberty, e.g., periungual fibromas. It seems possible that either hormonal factors or growth factors may influence the size of cardiac rhabdomyomas, since these lesions may be large in newborns and then decrease in size.

Studies Aimed at Determining the Basic Genetic Defect in TSC

To date no consistent protein or enzyme abnormality has been identified in TSC. In the past few years a number of investigators have initiated genetic linkage studies in TSC for the purpose of mapping the TSC gene or genes to a specific chromosomal region. It will then be possible to apply the so-called reverse genetics approach to isolating the TSC gene. This approach has been applied successfully to the isolation of the cystic fibrosis gene. In addition, an effort has been made to identify individuals with TSC and dysmorphic features suggestive of a chromosomal abnormality, since the presence of a chromosomal deletion or translocation in a sporadic case of TSC may indicate a chromosomal region that is important in the pathogenesis of TSC. Chromosomal deletions and translocations have greatly expedited the regional mapping and disease-gene isolation in Duchenne muscular dystrophy and neurofibromatosis.

Genetic Linkage Studies in TSC

In 1987, Fryer et al. published evidence for linkage of TSC and the ABO blood group locus on human chromosome 9q34. This evidence was a peak cumulative lod score of 3.85 at $\theta = 0$, between TSC and *ABO*. Connor et al. (1987) reported results of linkage analyses in three Scottish families that revealed a peak lod score of 3.18 at $\theta = 0$, between TSC and another chromosome 9q34 marker, *ABL*.

Following these initial reports, there were a number of reports from other investigators of linkage analysis data that did not support linkage of TSC and the chromosome 9q34 region. Results of two-point and multipoint analysis of TSC and three chromosome 9 markers, *ABO, ABLK2,* and *MCT136*, by Kandt et al. (1989) led them to conclude that location of the TSC locus could be excluded for a distance of 20 cM adjacent to the ABO locus.

In 1990, Smith et al. published results of two-point and multipoint linkage analysis in 15 TSC families. In the pairwise linkage analysis, using a penetrance value of 90%, a significant positive lod score was observed between TSC and the chromosome 11q22–11q23 marker *MCT128.1* (*D11S144*): 3.26 at θ = 0.08. A probe for tyrosinase (mapped in the 11q14–q22 region), gave a maximum lod score of 2.88 at θ = 0. Results of multipoint analysis indicated that the most likely order was (*TYR TSC*)—*MCT128.1*–*HHH172*.

Janssen et al. (1990) investigated nine multigenerational families with TSC. For linkage analysis they utilized an approach that combined multipoint linkage analysis and heterogeneity testing. Their results supported a model with two different loci determining TSC. Results of their studies mapped the *TSC1* locus in the vicinity of the *ABL* locus on chromosome 9q34 and supported assignment of a *TSC2* locus in the interval between the *D11S29* locus and the locus for the dopamine D2 receptor, on chromosome 11.

Analysis of Genetic Heterogeneity in TSC

A number of investigators have recently pooled their linkage data obtained in TSC families so that a large data set may be available for analysis of genetic linkage and linkage heterogeneity. Haines et al. (1990) analyzed 111 families in a collaborative data set. They used the LINKAGE and LINKMAP programs to determine two-point lod score and multipoint location scores. In addition, they used the HOMOG programs to test for homogeneity of linkage. Using the chromosome 9 data, they obtained highly significant evidence for rejection of homogeneity (χ^2 = 21.54, p = 0.0001). When the 9q and 11q data were used simultaneously, the analysis again rejected homogeneity (χ^2 = 39.74, p = .0001). They also examined the combined data for evidence of a third locus; results of this analysis were suggestive but not significant. Haines et al. (1990) determined that the maximum likelihood estimate of the proportion of chromosome 9–linked families was 0.38, of chromosome 11–linked families was 0.47, and of unlinked families was 0.15. Results of multipoint analysis revealed a location score for chromosome 9q of 6.51. In the case of chromosome 11q, the peak location score was 2.77. The most likely position of the 9q TSC gene was near the ABL oncogene.

Future Prospects

From the studies described above it is clear that there are at least two TSC-determining genes, one located on chromosome 9q34 and the other located possibly in the chromosome 11q22–11q23 region. For further progress in linkage mapping it will be important to identify additional highly polymorphic markers in these gene regions and to design analyses that will allow one to clearly determine whether a particular TSC family represents a 9-linked or an 11-linked family.

Further progress is also dependent on the development of more accurate maps of chromosome 9 and 11 markers, since the power of multipoint linkage analyses is greatly diminished by inaccuracies in the map relationships of genes. Linkage maps of chromosome 9 and of chromosome 11 have been published, and these maps will provide a basis for further expansion. Linkage studies using markers on chromosomes other than 9 and 11 will be necessary given the suggestion in certain analyses that there may be a third TSC locus.

SELECTED REFERENCES

Baraitser M, Patton MA. Reduced penetrance in tuberous sclerosis. J Med Genet 1985;22:29–31.

Bourneville D-M. Sclérose tubéreuse des circonvolutions cérébrales: idiotie et épilepsie hémiplégique. Arch Neurol (Paris) 1880;1:81–91.

Bourneville D-M. Encéphalite ou sclérose tubéreuse des circonvolutions cérébrales. Arch Neurol (Paris) 1881;1:390–412.

Cavenee WK, Dryja TP, Phillips RA. Expression of recessive alleles by chromosomal mechanisms in retinoblastoma. Nature 1983;305:779–784.

Connor JM, Pirritt LA, Yates J, Fryer EA, Ferguson-Smith MA. Linkage of tuberous sclerosis to DNA polymorphism detected by vABL. J Med Genet 1987;24:544–546.

Fryer AE, Chalmers A, Connor JM, et al. Evidence that the gene for tuberous sclerosis is on chromosome 9. Lancet 1987;1:659–661.

Gomez MR. History. In: Gomez MR, ed. Tuberous sclerosis. New York: Raven Press, 1988.

Gomez MR. Criteria for diagnosis. In: Gomez MR, ed. Tuberous sclerosis. 2nd ed. New York: Raven Press, 1988:9–19.

Gomez MR. Neurologic and psychiatric features. In: Gomez MR, ed. Tuberous sclerosis. 2nd ed. New York: Raven Press, 1988.

Gunther M, Penrose LS. The genetics of epiloia. J Genet 1935;31:413–430.

Haines J, and the Tuberous Sclerosis Collaborative Group. Genetic heterogeneity in tuberous sclerosis: study of a large collaborative dataset. Tuberous sclerosis and allied diseases. Ann N Y Acad Sci 1990;615:256–264.

Houser OW, Nixon JR. Central nervous system imaging. In: Gomez MR, ed. Tuberous sclerosis. 2nd ed. New York: Raven Press, 1988.

Huttenlocher PR, Heydeman PT. Fine structure of cortical tubers in tuberous sclerosis. Ann Neurol 1984;16:595–602.

Janssen LAJ, Sandkuyl LA, Merkens EC, et al. Genetic heterogeneity in tuberous sclerosis. Genomics 1990;8:237–242.

Julier C, Nakamura Y, Lathrop M, et al. A detailed map of the longarm of human chromosome 11. Genomics 1990;7:335–345.

Kandt RS, Pericak-Vance M, Hung W, et al. Absence of linkage of ABO bloodgroup locus to familial tuberous sclerosis. Exp Neurol 1989;104:223–228.

Kunkel LM, Monaco AP, Middlesworth W, Ochs HD, Latt S. Specific cloning of DNA fragments absent from the DNA of a male patient with an X chromosomal deletion. Proc Natl Acad Sci USA 1985;82:4778–4782.

Lathrop M, Nakamura Y, O'Connell P, et al. A mapped set of genetic markers for human chromosome 9. Genomics 1988;3:361–366.

Michel JM, Diggle JH, Brice J, et al. Two half-siblings with tuberous sclerosis, polycystic kidneys and hypertension. Dev Med Child Neurol 1983;25:239–244.

Nixon JR, Miller GM, Okazaki H, Gomez MR. Cerebral tuberous sclerosis: postmortem magnetic resonance imaging and pathologic anatomy. Mayo Clin Proc 1989;64:305–311.

Reagan TJ. Neuropathology. In: Gomez MR, ed. Tuberous sclerosis. 2nd ed. New York: Raven Press, 1988:63–74.

Rommens J, Ianuzzi M, Kerem B, et al. Identification of the cystic fibrosis gene: chromosome walking and jumping. Science 1989;245:1059–1065.

Shepherd CW, Gomez MR, Lie JT, et al. Causes of death in patients with tuberous sclerosis. Mayo Clin Proc 1991;66:792–796.

Shepherd CW, Scheithauer BW, Gomez MR, Altermatt HJ, Katzmann JA. Subependymal giant cell astrocytoma. Neurosurgery 1991;28:864–868.

Smith M, Smalley S, Cantor R, et al. Mapping of a gene determining tuberous sclerosis to human chromosome 11q14–11q23. Genomics 1990;6:105–114.

Van der Hoeve J. Eye symptoms in tuberous sclerosis of the brain. Trans Ophthalmol Soc UK 1920;40:329–334.

Viskochil D, Buchberg A, Xu G, et al. Deletions and a translocation interrupt a cloned gene at the neurofibromatosis type I locus. Cell 1990;62:187–192.

Vogt H. Zur Diagnostik der Tuberösen Sklerose. Z Erforsch Behandl Jugendl. Schwachsinns 1908;2:1–12.

von Recklinghausen F. Ein Herz von einem Neugeborenen welches mehrere theils nach ausen, theils nach den Holden prominirende tumoren (Myomen) trug. Verh Ges Geburtsh 1862;20:1–2.

Wiederholt WC, Gomez MR, Kurland LT. Incidence and prevalence of tuberous sclerosis in Rochester, Minnesota, 1950 through 1982. Neurology 1985;35: 600–603.

Xu GF, O'Connell P, et al. The neurofibromatosis type 1 gene encodes a protein related to GAP. Cell 1990;62:193–201.

CHAPTER 11

Charcot-Marie-Tooth Disease

Jeffery Vance
Tom Bird

The Charcot-Marie-Tooth (CMT) hereditary neuropathy syndrome is one of the most heterogeneous of the inherited neurologic disorders, demonstrating both genetic and clinical heterogeneity. Recent linkage studies in the autosomal dominant and X-linked subtypes have made this aspect of the disease even more intriguing. Other late-onset dominant neurologic disorders such as Huntington's disease, myotonic dystrophy, and neurofibromatosis, have not shown similar genetic variability. This genetic heterogeneity is being elucidated primarily through the use of pedigree linkage analysis. CMT is a model for the power of this methodological approach in classifying such a group of previously confusing disorders.

The presence of multiple genetic defects in CMT provides a series of windows into the basic processes of peripheral nerve structure and function. As the genetic defects of the peripheral nerve are clarified, new insights should be gained concerning the complex mechanisms of nerve development and action.

History and Clinical Features

Hereditary progressive muscular atrophy in the lower extremities, with onset in the first decade, was described by Charcot and Marie in 1886, and independently in the same year by Tooth in his physician thesis. The original summary by Charcot and Marie of the primary clinical features of this disorder, is probably still applicable today:

> Progressive muscular atrophy, first invading the feet and legs and only appearing in the upper limbs (hands first, then the forearms) several years later; thus, slow evolution. Relative integrity of the proximal muscles of the limbs, or at least much longer preservation than the

distal muscles. Integrity of the muscles of the trunk, shoulders, and face. Existence of fibrillary contractions in the muscles undergoing atrophy. Vasomotor disturbances in the affected segments of limbs. No notable tendinous contractures on the side of articulations where the muscles are atrophied. Sensation most often intact, though sometimes altered in various ways. Frequency of cramps. Reaction of degeneration in the muscles undergoing atrophy. Beginning of the disease usually during childhood, often among several brothers and sisters; sometimes it exists not only in collateral relatives but also in the forebears.

Tooth suggested this form be termed the "peroneal type of progressive muscular atrophy" and noted that based on autopsies it was probable that it was a "true neuropathy." Charcot and Marie, noting the distal onset of the disease, felt it unlikely to be a myopathy. Rather they believed that "up to a certain point the hypothesis of a myelopathy appears preferable . . . but it seems difficult to make an absolute statement."

Today the Charcot-Marie-Tooth syndrome is known to be the most common of the inherited neuropathies, estimated to have an occurrence of approximately 36 per 100,000 for the most common type, the autosomal dominant form. Onset is most common in childhood or in young adulthood. Weakness begins in the peroneal nerve distribution, with development of the characteristic foot drop. The high arch, or pes cavus, as well as hammer toes often accompany the progressive distal atrophy and weakness of the lower leg. This produces the stork leg or inverted bottle appearance of the foot and calf noted by many authors. While the sensory examination is usually abnormal, the main symptoms are primarily motor. As the disease progresses it may begin to affect the strength of the hands as well, leading to muscle atrophy similar to that in the lower extremities. The severity of the disease cannot be predicted in any one individual. Some patients are severely incapacitated, requiring extensive orthotic support. Others may exhibit only minimal symptoms, requiring either nerve conduction studies (if applicable) or obligate carrier status for diagnosis.

An action tremor and ataxia can be found in individuals with CMT. Some authors describe the tremor as similar to benign essential tremor, and more common in CMT type 1 than type 2. At times, these symptoms may be prominent and confuse the diagnosis of CMT in some families (see next section).

MENDELIAN HERITABILITY

CMT is characterized by known genetic heterogeneity. The most common pattern of inheritance is as an autosomal dominant disorder, although X-linked and recessive forms have been reported. In 1968, Dyck and Lambert divided the autosomal dominant forms into two types, based on pathologic and physiologic criteria: (1) the demyelinating type (CMT type 1), with significantly decreased nerve conduction velocities (NCV) and hypertrophic changes on nerve biopsy with hyperplasia of Schwann cells, termed onion bulb formation; and (2) the axonal or neuronal form (CMT type 2), in which families presented with normal or slightly slowed NCV and without the hypertropic changes on nerve biopsy. These two types are for all practical purposes clinically indistinguishable. Later, Thomas et al. confirmed these findings and introduced the term hereditary motor and sensory neuropathies (HMSN) to describe the inherited peroneal atrophies. Seven HMSN types have been classified by Dyck using pathologic, physiologic, and clinical differences. CMT type 1 is referred to as HMSN I and CMT type 2 as HMSN II (Table 11.1).

The possibility of an X-linked form of CMT was originally raised by Herringham in 1888, in an interesting and insightful description:

> This form makes one wonder what inheritance is. That the diseased tissues of a consumptive father should be so represented in his spermatozoon as to cause his child to fall into consumption is remarkable enough. But that the women of this family, themselves even uncommonly buxom and healthy, should be able to select their children, and transmit to the males alone tissues unlike their own, and endowed with

Table 11.1: The Charcot-Marie-Tooth Syndromes

Type	Pathophysiology	Chromosome Location
CMT 1 (HMSN I)	Demyelinating, with hypertrophic changes on nerve biopsy, onion bulb formation, and decreased nerve conduction velocities.	
Autosomal dominant		
CMT 1A		17p.11.2
CMT 1B		1q
CMT 2 (HMSN II)	Neuronal form with normal or mildly slowed nerve conduction velocities, no hypertrophic changes on nerve biopsy.	Unknown
CMT X-linked		Xq11–q13
HMSN III		
Autosomal recessive		Unknown

a regular form of weakness which they do not themselves possess, is still, more marvelous. It seems as if the daughter of a diseased father carried from the beginning of her life ova of two sexes, the female healthy, the male containing within it the representation of the father's disease.

In 1939, Allan first suggested X linkage in a North Carolina family. However, Thomas felt that this and other families were not X-linked but examples of variable expression between sexes, with females more likely to have "minor expression." This debate continued until Allan's subject family was further investigated by Rozear et al. In extending this family, no male-to-male transmission was observed in 25 opportunities. This provides odds of 32 million to one for X-linked inheritance. In addition, linkage was established to X-chromosome markers (see below), confirming X-linked inheritance. This case illustrates the use of linkage analysis and molecular genetics to demonstrate a major gene locus among a clinically homogeneous group of families.

Several additional families have now exhibited X linkage, with the frequency of this form estimated to be 3.6 per 100,000. Most families with the X-linked disorder display the primarily demyelinating type of CMT, although recently an axonal X-linked form of CMT has been proposed. In general, phenotypic expression appears to be similar to that of the autosomal dominant form. However, males are much more severely affected than females. NCVs are more variable, with absent or severely decreased responses in males, while obligate carrier females ranged from 5 to 49 m/sec (normal values are ≥ 41 m/sec). In addition, some obligate carriers were asymptomatic, with normal NCVs. X-linked recessive and X-linked dominant families have been previously described. Whether these are examples of variable clinical expression of the same disorder or distinct genetic entities remains to be demonstrated.

Dejernne-Sotas syndrome (DS) or HMSN III is an autosomal recessive syndrome in which onset is in infancy and symptoms are quite severe, with NCV usually less than 10 m/sec. But apparent autosomal recessive families with CMT have been reported that would appear to be clearly distinct from DS. In these reported autosomal recessive families, multiple siblings are affected and have "normal" parents, although some parents have not had NCV tests performed. Results have been normal in those parents who have had NCV tests. Consanguinity is frequently reported, lending support to recessive inheritance. The presence of multiple affected siblings and consanguinity makes nonpaternity unlikely, but should be considered. Some reported families have had significantly low nerve conduction velocities, while others have had normal or only slightly decreased NCVs. The phenotypes of these recessive families have been similar to the dominant types, but the average patient's symptoms are generally more severe. Only one segregation study of a presumed autosomal recessive family has been reported, supporting autosomal recessive inheritance. The estimated incidence of autosomal recessive CMT is approximately 1.4 per 100,000.

In 1926, Roussy and Levy described a family with peroneal atrophy in which the cases occurred with ataxia or tremor. Onset was early in life, and these subjects were initially felt to be expressing a distinct clinical entity. However, subsequent families have been observed in which some individuals present only with CMT 1, but where additional members of the family have features similar to those described by Roussy and Levy. Therefore, it is now felt that the Roussy and Levy syndrome is an example of variable expressivity in CMT 1 and not a distinct genetic entity. Linkage studies in "Roussy-Levy" families could prove a useful tool with which to delineate the validity of this clinical impression.

Penetrance in the dominant forms of CMT is believed to be 100%, but with variable age of onset and expressivity. The available data suggest that, on average, CMT 2 patients have a later age of symptom onset than type 1 patients, but with considerable overlap between the two types. The mean age of onset of clinical symptoms in type 1 is in the second decade, but the range of onset has been reported to be from several years of age to more than 50 years of age. With the initiation of family studies, several individuals, now well into their 30s, have been documented to be carriers based upon slowed NCV, but remain asymptomatic on exam. Also, estimating age of onset may be a difficult task. Certainly many milder cases go unreported or do not come to attention until late in life, when the neuropathy begins to interfere with function.

Electrical diagnosis in type 1 does not demonstrate such variability in age of onset. Abnormal NCVs have been observed in infants at six months of age. However, NCVs do not reach adult levels until the age of 5–7 years. Therefore, we would suggest that these ages are a more appropriate time for carrier testing at most centers. Carrier testing done prior to this age should involve an age-matched control population. Gene carriers will demonstrate slow NCV, usually 60% or less of normal, often years before they exhibit any symptoms. CMT 1 families are therefore quite useful in linkage analysis, as most young individuals' electrical tests are informative concerning their affected status.

LINKAGE STUDIES

In 1980, Bird et al. first suggested linkage of CMT to the Duffy blood group locus, which is located in the pericentromeric region of chromosome 1 at

1q21.1–1q23.3. Bird et al.'s maximum lod score (z) was 2.30 at $\theta = 10$ centimorgans (cM). Shortly thereafter, Guilloff, using two families, added a score of $z = 0.725$ to raise the cumulative lod score to $z = 3.025$ at 10 cM. An independent Indiana family was subsequently reported by Stebbins and Conneally to be linked to the Duffy locus with $z(\theta) = 3.11$. Additional families were collected with the expectations of similar linkage findings to chromosome 1. However, it soon became clear that the Duffy linkage was not as definitive as initially expected. Subsequent studies demonstrated that many families were clearly not Duffy linked. This led Bird et al. to suggest that the non–Duffy-linked types be termed CMT 1A, while those demonstrating linkage to the Duffy blood group be designated type 1B.

While many investigators studied additional markers on chromosome 1 to make existing families more informative, a primary difficulty was in the evaluation of positive lod scores in the presence of known genetic heterogeneity. Small positive lod scores from a nonlinked family commonly occur by chance. But if the lod score value is less than 3.0, conventionally accepted to be stringent enough to be linked, what are the chances of linkage of that family to the marker? This chance can be estimated by considering the prior probability of linkage to that marker for any one family weighted by the lod score obtained for that family. The prior probability of linkage is the estimated percentage of families linked to that marker in the studied population.

However, the actual proportion of families linked to the Duffy locus was unknown. The failure to deal correctly with the inherent problem of heterogeneity led to several reports of Duffy linkage in selected families with only slightly positive scores, while those with negative values were stated to be nonlinked. In fact, most of the small positive lod scores were most likely the result of chance and not indicative of linkage to the Duffy locus. This problem has been well discussed by Ott, with particular reference to CMT. For those individuals not wishing to calculate the posterior probability by hand, or if the prior probability is not known, Ott's program HOMOG can be used.

Using four Duke families together with two families from the University of Sydney, Vance et al. reported linkage between the two chromosome 17 markers, *D17S58* and *D17S71*, and CMT 1A. Confirming studies by several investigators in Europe and the United States have demonstrated that the majority of large type-1 families presently studied are linked to these markers on chromosome 17. Middleton-Price et al. demonstrated linkage to chromosome 17 in their family studies, which included the original two families used in Guilloff's earlier Duffy linkage report. While one of these two families was uninformative, the other family had a score of $z = 1.65$ at $\theta = 0.05$ with *D17S71*. Given that the prior probability of linkage to chromo-

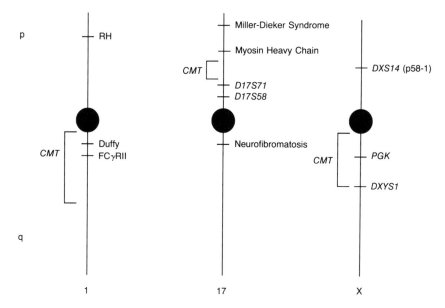

Figure 11.1: Location of known loci for CMT 1.

some 17 appears high for type 1 families, it would appear that the initial positive score reported by Guilloff et al. was most likely a chance event.

While some families originally believed to be linked to chromosome 1 now demonstrate a much higher probability of chromosome 17 linkage, the original type 1B family of Stebbins and Conneally has continued to be linked to the Duffy region of chromosome 1, by the typing of additional markers in that region. Lebo et al. have reported linkage in this Indiana family to the IgG receptor FC-gamma-RII, which they report is located within 6 cM of the Duffy locus. Unfortunately, Bird et al.'s original family has not been informative for this marker, but has been excluded from the chromosome 17–linked region and therefore is most likely also a chromosome 1 family.

Vance et al. have recently demonstrated conclusively (with odds > 100,000:1) that the chromosome 17 locus lies on the p arm. The method used to sublocalize the CMT gene was multipoint linkage analysis. This method involves performing linkage analysis with several linked markers at once, allowing their combined information to estimate the most likely marker-to-marker interval that contains the CMT gene. Currently, the CMT 1A gene is most likely flanked by the markers *D17S122* and *D17S125* (a distance of 3 cM) and appears to lie proximal to the myosin heavy chain locus.

Few linkage studies have been reported in CMT 2. With the recent linkage of type 1A to chromosome 17, Loprest et al. have excluded linkage in a large type 2 family to the chromosome 17 and chromosome 1 region loci. This suggests that CMT 2 is genetically as well as pathologically distinct from the known linked type 1 loci.

The use of chromosomal abnormalities would be of great use in localizing the CMT 1A gene. However, to date, none of the individuals studied have demonstrated the localized structural chromosomal abnormalities (translocations) seen in a few patients with neurofibromatosis. If such an individual could be identified, with a distinct localized chromosomal abnormality, this would greatly hasten the mapping of the CMT gene.

Recently, however, Lupski et al. and Raeymaekers et al. have reported the presence of a chromosomal duplication in CMT 1A patients for the probe VAW 409 R1 (*D17S122*). Using a 409 R1-derived CA repeat and fluorescent in situ hybridization, Lupski et al. demonstrated the presence of three alleles in CMT 1A patients, indicating a duplication (two alleles) on one chromosome. An abnormal SAC II fragment was also seen in patients by using pulsed field gel electrophoresis. Concurrently, Raeymaekers et al. also reported the duplication, observing increased allele (band) density in CMT 1A patients on Southern blots as well as abnormal fragments on pulsed field electrophoresis. Vance et al. have subsequently supported their finding, demonstrating the segregation of three alleles for two of their six known CMT 1A families using the 409 R1 CA repeat. The size of this duplication and its frequency in CMT 1A families is as yet unknown, although it has been postulated to be as large as a megabase. It appears to suggest a mechanism for the disorder in most families. These findings should facilitate identification of the disease locus on chromosome 17. The presence of a duplication also suggests that CMT 1A may be the result of a unique mechanism, a segmental trisomy.

The heterogeneity observed in type 1 may exist within type 2 pedigrees as well. Families with biopsies consistent with type 2 have been reported with "intermediate" NCVs, i.e., NCVs in which the range within the family includes values lower than classic type 2 values but higher than those expected for type 1. Whether these are intermediate families or these families are exhibiting natural variability of NCV will remain unanswered until linkage analysis can be used to identify a major locus. As linkage analysis defines subgroups of hereditary neuropathy, parallel clinical studies will, one hopes, carefully search for phenotypic differences between the groups.

X-Linked Families

Initial analysis in X-linked families revealed evidence of linkage to the X chromosome marker *DXYS1,* located on the proximal long arm of the X chromosome. Multipoint analysis has determined the most likely location of the gene to be between the centromere and *PGK,* in the pericentromeric region of the *q* arm. Recently, an additional large Canadian family has demonstrated lod scores suggestive of this region as well.

ADDITIONAL APPROACHES

The biochemical defect in CMT has been variously postulated to be within the Schwann cell or in the axon itself. No abnormality of the major myelin proteins has been demonstrated and it appears unlikely that they are candidate genes for CMT. Indeed, many of these proteins have been located on either the human or the mouse genome and do not correspond to the known CMT regions.

No exact animal models exist for CMT. However, the Trembler mouse (Tr) is a possible candidate. The Tr mouse develops a hypomyelinating peripheral neuropathy with an autosomal dominant inheritance pattern. While hypomyelination is prominent in early age, demyelination and remyelination with onion bulb formation have been described in older animals.

The Tr locus is known to be located just distal to the vestigial tail locus on mouse chromosome 11. This is intriguing because of the high homology between mouse chromosome 11 and human chromosome 17. The defect has recently been shown to be a point mutation in one of the Schwann cell "growth arrest–specific" genes, GAS 3. It is not currently known if the Tr locus lies within the region homologous to the chromosome 17 loci linked to CMT, but work is currently underway to investigate this candidate gene in CMT 1A.

GENETIC COUNSELING

The counseling of a disorder as heterogeneous as CMT should always proceed with caution. Initially, the more common etiologies (diabetes, alcohol, toxins) for the patient's neuropathy need to be ruled out. Written or physical documentation of a similar disease in the patient's family must be acquired, as verbal reports can be misleading. We have frequently found that additional affected family members are located among relatives of individuals who thought of themselves as an isolated case. Other inherited neuropathies that include the CMT phenotype as part of their syndrome of findings must also be considered.

The determination of the type to which any single family belongs relies mainly on its inheritance pattern, and physiologic and linkage data. For example, unless male-to-male transmission is observed, it may be difficult to rule out X-linked inheritance in many families without linkage data or a large family structure. The finding of unaffected parents should always raise the consideration of nonpaternity. Nerve biopsies may be helpful, but are not specific. Nerve conduction tests should be performed on all at-risk individuals prior to any diagnosis, as mild physical findings may be misleading. In fact, individuals well in their 30s without any signs of neuropathy, but who are carriers of the type 1 gene by NCV tests or family studies, have been documented.

Linkage analysis will eventually prove to be one of the most useful of these methods of diagnosis. However, at this time the knowledge concerning linkage relationships cannot be used for diagnosis or genetic counseling in CMT. This is because the proportion and degree of genetic heterogeneity is not yet complete, and exact distances of markers from the CMT gene are not clear. With additional probes and further investigation, this situation may change in the not-too-distant future. Finally, the presence of an abnormal SAC II fragment on pulsed field gel electrophoresis in most CMT 1A patients may provide a method of carrier detection as well.

SUMMARY

The CMT syndrome is proving to be a fascinating disorder with extensive genetic heterogeneity. At least four distinct genetic types of the demyelinating form (CMT 1A, CMT 1B, autosomal recessive, and X-linked CMT) are known. In addition, the neuronal type, CMT 2 (HMSN II), demonstrates similar diversity. These distinct lesions are likely to involve multiple events in a major pathway or pathways leading to peripheral nerve growth, myelination, and repair.

Elucidation of the defects should provide insight into these mechanisms. In addition, at least in the demyelinating types, gene carriers have signs of the disease years before symptoms develop, so identification of the genetic defect may lead to a preventive therapy that can be used in this asymptomatic interval. Therefore, this disorder promises to be a fruitful arena for the investigations of both genetic heterogeneity and peripheral nerve function.

SELECTED REFERENCES

Allan W. Relation of hereditary pattern to clinical severity as illustrated by peroneal atrophy. Arch Intern Med 1939;63:1123–1131.

Bird TD, Ott J, Giblett ER. Linkage of Charcot-Marie-Tooth neuropathy to the Duffy locus on chromosome 1 (Abstract). Am J Hum Genet 1980;32:99A.

Bird TD, Ott J, Giblett ER, Chance PF, Sumi SM, Kraft GH. Genetic linkage evidence for heterogeneity in Charcot-Marie-Tooth neuropathy (HMSN type I). Ann Neurol 1983;14:679–684.

Brody IA, Wilkins RH. Neurologic classics III. Charcot-Marie-Tooth disease. Arch Neurol 1967;17:552–557.

Buchberg AM, Brownell E, Nagata S, Jenkins NA, Copeland NG. A comprehensive genetic map of murine chromosome 11 reveals extensive linkage conservation between mouse and human. Genetics 1989;122:153–161.

Chance PF, Bird TD, O'Connell P, Lipe H, Lalouel J-M, Leppert M. Genetic linkage and heterogeneity in type I Charcot-Marie-Tooth disease (hereditary motor and sensory neuropathy type I). Am J Hum Genet 1990;47:915–925.

Charcot JM, Marie P. Sur une forme particuliere d'atrophie musculaire progressive souvent familiale debutant par les preds et les jambes et attergnant plus tord les mains. Revue de Medecine 1886;6:97–138.

Dyck PJ. Inherited neuronal degeneration and atrophy. In: Dyck PJ, ed. Diseases of the peripheral nervous system. 2nd ed. Philadelphia: W.B. Saunders Company, 1984:1609–1630.

Dyck PJ, Lambert EH. Lower motor and primary sensory neuron diseases with peroneal muscular atrophy. II. Neurologic, genetic and electrophysiologic findings in various neuronal degenerations. Arch Neurol 1968;18:619–625.

Fischbeck KH, ar-Rushdi N, Pericak-Vance MA, Rozear M, Roses AD, Fryns JP. X-linked neuropathy: gene localization with DNA probes. Ann Neurol 1986;20:527–532.

Fischbeck KH, Ritter A, Shi Y-J, et al. X-linked recessive and X-linked dominant Charcot-Marie-Tooth disease. In: Lovelace RE, Shapiro HK, eds. Charcot-Marie-Tooth disorders: pathophysiology, molecular genetics, and therapy. New York: John Wiley & Sons, 1989.

Guilloff RJ, Thomas PK, Contreras M, Armitage S, Schwarz P, Sedgwick EM. Evidence for linkage of type I hereditary motor and sensory neuropathy to the Duffy locus on chromosome 1. Ann Hum Genet 1982;46:25–27.

Gutmann L, Fakadej A, Riggs JE. Evolution of nerve conduction abnormalities in children with dominant hyperprophic neuropathy of the Charcot-Marie-Tooth type. Muscle Nerve 1983;6:515–519.

Hahn AF, Brown WF, Koopman WJ, Feasby TE. X-linked dominant hereditary motor and sensory neuropathy. Brain 1990;113:1511–1525.

Harding AE, Thomas PK. Genetic aspects of hereditary motor and sensory neuropathy (types I and II). J Med Genet 1980;17:329–336.

Herringham WP. Muscular atrophy of the peroneal type affecting many members of a family. Brain 1888;11:230–236.

Ionasescu V, Murray JC, Burns T, Ionasescu R, Ferrell R, Searby C. Linkage analysis of Charcot-Marie-Tooth neuropathy (HMSN type I). J Neurol Sci 1987;80:73–78.

Lebo RV, Dyck PJ, Chance PF, et al. Charcot-Marie-Tooth locus in FC-gamma-RII gene region (Abstract). Am J Hum Genet 1989;45:A148.

Loprest L, Vance JM, Pericak-Vance MA, et al. Linkage studies demonstrate

Charcot-Marie-Tooth disease types 1 and 2 are distinct genetic entities. Neurology 1992 (in press).

Lupski JR, Montes de Oca-Luna R, Slaugenhaupt S, et al. DNA duplication associated with Charcot-Marie-Tooth disease type 1A. Cell 1991;66:219–232.

Middleton-Price HR, Harding AE, Monteiro C, Berciano J, Malcolm S. Linkage of hereditary motor and sensory neuropathy type I to the pericentromeric region of chromosome 17. Am J Hum Genet 1990;46:92–94.

Ott J. Analysis of human genetic linkage. Baltimore: The Johns Hopkins University Press, 1985:105.

Raeymaekers P, Timmerman V, Nelis E, et al. Duplication in chromosome 17p11.2 in Charcot-Marie-Tooth neuropathy type 1A (CMT 1A). Neuromusc Disord 1991;51:93–97.

Roussy G, Levy G. Sept cas d'une maladie familiale particuliere: trouble de la marche, preds bots et areflexie tendineuse generalisee, avec accessoirement, legere maladresse des mains. Rev Neurol 1926;1:427–450.

Rozear MP, Pericak-Vance MA, Fischbeck K, et al. Hereditary motor and sensory neuropathy, X-linked: a half century follow-up. Neurology 1987;37: 1460–1465.

Stebbins NB, Conneally PM. Linkage of dominantly inherited Charcot-Marie-Tooth neuropathy to the Duffy locus in an Indiana family (Abstract). Am J Hum Genet 1982;34:195A.

Suter U, Welcher AA, Ozcelik TA, et al. The trembler mouse carries a point mutation in a myelin gene. Nature 1992 (in press).

Thomas PK, Calne DB, Stewart G. Hereditary motor and sensory polyneuropathy (peroneal muscular atrophy). Ann Hum Genet 1974;38:111–153.

Tooth HH. Recent observations on progressive muscular atrophy. Brain 1887;10:243–253.

Vance JM, Barker D, Yamaoka LH, et al. Localization of Charcot-Marie-Tooth disease type 1a (CMT 1a) to chromosome 17p11.2. Genomics 1991;9:623–628.

Vance JM, Nicholson GA, Yamaoka LH, et al. Linkage of Charcot-Marie-Tooth neuropathy type 1a to chromosome 17. Exp Neurol 1989;104:186–189.

Vance JM, Pericak-Vance MA, Lucas A, et al. Genetic mapping of the CMT 1A locus (Abstract). Hum Gene Map 11 (in press).

CHAPTER 12

Mitochondrial Myopathies

A. E. Harding
I. J. Holt

THE MITOCHONDRIAL GENOME

Mammalian mitochondria contain 2 to 10 circular DNA molecules that are double-stranded and about 16.6 kilobases (kb) in length, contributing about 1% of total cellular DNA. Human mitochondrial DNA (mtDNA) has been sequenced and been found to differ from nuclear DNA to some extent in its genetic code and also in that it contains very little noncoding sequence. Each strand of mtDNA is transcribed from a single promotor site and then processed. The heavy (H) strand transcripts consist of 2 ribosomal RNAs, 14 tRNAs, and 12 protein-coding sequences, and the light (L) strand codes for eight tRNAs and 1 protein-coding sequence. The mitochondrial protein-coding transcripts are not capped at their 5′ end, but their 3′ ends are polyadenylated by mitochondrial poly(A) polymerase. Mitochondria generally divide at a rate appropriate to that of division of their parent cells, and in most instances mtDNA molecules also replicate with every cell cycle.

mtDNA encodes for 13 of the 67 or so subunits of the mitochondrial respiratory chain and oxidative phosphorylation system: seven subunits of complex I (NADH dehydrogenase; ND), cytochrome b (complex III); subunits I, II, and III of cytochrome oxidase (complex IV); and subunits 6 and 8 of ATP synthetase. The nuclear genome encodes the remaining polypeptides in the respiratory chain and also controls their transport into mitochondria by synthesizing leader peptides, which appear to direct the proteins to sites of adhesion between the inner and outer mitochondrial membranes prior to transport into the matrix. Replication, transcription, and translation of the mitochondrial genome are also dependent on nuclear products such as the enzymes involved in mtDNA replication and transcription, and ribosomal proteins.

A few paternal mitochondria may penetrate the ovum at the time of fertilization, but these appear to degenerate subsequently, and mtDNA is effectively exclusively maternally transmitted in many species, including man. mtDNA from all individuals in a single maternal line shows the same pattern of fragments after digestion with restriction endonucleases, but extensive mtDNA nucleotide sequence divergence occurs between different maternal lines. The mutation rate of mtDNA is high; restriction mapping of mtDNA in different human populations can be used to trace their origins. Despite its high mutation rate, it is generally thought that all the mtDNA in an individual normal human is identical, although extensive sequencing studies of mtDNA from different tissues have not been described. mtDNA heteroplasmy (the presence of two populations defined by sequence variation) has been demonstrated in Drosophila and a single maternal line of Holstein cows.

In recent years, a number of human diseases have been associated with defects, or possible defects, of the mitochondrial genome that are often heteroplasmic (Figure 12.1). These diseases form the subject of this chapter, and are summarized in Table 12.1.

Table 12.1 Genetic Defects in Mitochondrial Diseases

Disease	Genetic Defect(s)
Mitochondrial myopathies	
PEO, KSS	Large mtDNA deletions, mtDNA duplications
MERRF	Point mutation of mtDNA
MELAS, other encephalomyopathies	Point mutations of mtDNA
Myopathy and cardiomyopathy	Point mutation of mtDNA
Not known	Possible defects of nuclear-encoded respiratory chain subunit genes
PEO	Multiple mtDNA deletions 2° to nuclear defects
Pearson's syndrome	Large mtDNA deletions
Leber's hereditary optic neuropathy	Point mutation of mtDNA
NARP	Point mutations of mtDNA

PEO = progressive external ophthalmoplegia; KSS = Kearns-Sayre syndrome; MELAS = mitochondrial encephalomyopathy with lactic acidosis and stroke-like episodes; MERRF = myoclonic epilepsy with ragged red fibers; NARP = neurogenic muscle weakness, ataxia, and retinitis pigmentosa.

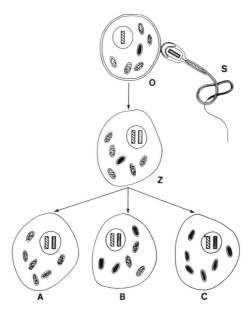

Figure 12.1: Cartoon (not to scale) illustrating maternal inheritance of mtDNA, compared with biparental inheritance of nuclear genes, and the random distribution of normal and mutant mitochondrial genomes in daughter cells of the zygote. It is assumed for simplicity that individual mitochondria contain either normal (open mitochondria) or mutant (filled mitochondria) mtDNA, not both. O = ooctye; S = sperm; Z = zygote; A, B, C = daughter cells of zygote, representing stem cells of different tissues. *Reprinted by permission from DiMauro S, et al. Clin Neurol 1990;8:494.*

MITOCHONDRIAL MYOPATHIES

Clinical Aspects

The term **mitochondrial myopathy** (MM) is applied to a clinically and biochemically **heterogeneous** group of diseases that share the feature of mitochondrial **structural abnormalities** in skeletal muscle. Ragged red fibers, containing **peripheral and** intermyofibrillar accumulations of abnormal mitochondria, **are seen** with the modified Gomori trichrome stain. These fibers **represent the** morphological hallmark of these disorders. Ragged red fibers **were initially** commonly observed in patients presenting with syndromes of **chronic progressive external ophthalmoplegia (CPEO)** and/or proximal myopathy, often with weakness induced or enhanced by exertion. More recently, morphological evidence of mitochondrial dysfunction in muscle has been described in children and adults with complex multisystem disorders predominantly or exclusively affecting the central nervous system (CNS). These disorders present with psychomotor retardation, dementia, ataxia, seizures, movement disorders, stroke-like episodes, pigmentary retinopathy, deafness, and peripheral neuropathy in various combinations. Involvement of the heart, endocrine system, kidney, and hemopoietic tissues has also been reported.

It has been suggested that cases of MM can be classified into distinct syndromes on clinical grounds. These include the Kearns-Sayre syndrome, a combination of CPEO and pigmentary retinopathy developing before the age of 20 years, associated with ataxia, cardiac conduction defects, and increased cerebrospinal fluid protein concentrations; the syndrome of mitochondrial encephalopathy, lactic acidosis, and stroke-like episodes (MELAS); and a further syndrome of myoclonus epilepsy with ragged red fibers (MERRF). These syndromes are not specific even on clinical grounds, as there is considerable overlap between them. They do represent combinations of some of the more striking features of the mitochondrial myopathies and are useful as pointers to the underlying diagnosis.

Biochemical Aspects

A defect of aerobic metabolism in muscle mitochondria was suggested by the observation that many patients with MM have a pathological increase in serum lactate concentration during and after exercise. Defects of the mitochondrial respiratory chain have been demonstrated in most cases investigated. NADH, a major product of oxidation of pyruvate and fatty acids in the mitochondrial matrix, is the principle substrate for the respiratory chain enzyme complexes that are embedded in the inner mitochondrial membrane. Most of the electron carriers in the chain contain metal atoms (heme or nonheme iron, and copper) that are bound to a protein surface. The NADH–coenzyme Q (CoQ) reductase complex (complex I) consists of about 26 polypeptides, including eight iron sulphur (FeS) proteins, and accepts electrons from NADH before transferring them to complex III (CoQ–cytochrome c reductase) via a small, lipid-soluble molecule, CoQ, also called ubiquinone. $FADH_2$ feeds into the chain at complex III via complex II (succinate–CoQ reductase). The 11 or so subunits of complex III include cytochromes b and c_1, and two FeS proteins. Electrons pass from ubiquinone to cytochrome c via complex III, and then to the cytochrome oxidase complex (complex IV), consisting of 13 subunits including cytochromes a and aa_3. Cytochrome oxidase forms two water molecules from O_2 and four electrons transferred from cytochrome c.

The electrochemical proton gradient created by electron transfer along the respiratory chain drives the production of ATP from ADP and phosphate by the enzyme complex ATP synthetase (complex V) bound to the inner mitochondrial membrane. The energy supplied by the gradient is also used to transport pyruvate and other mitochondrial enzyme substrates, and nuclear-encoded mitochondrial proteins, into the mitochondrial matrix.

Human muscle mitochondria can be isolated from large biopsy specimens in amounts sufficient for polarographic studies of oxidation and phosphorylation, and for the determination of cytochrome spectra. These studies have identified a variety of defects of the respiratory chain and oxidative phosphorylation system in patients with MM, usually involving complexes I, III, and IV, individually or in combination. All of these defects are associated with a wide range of clinical syndromes. Conversely, symptoms and signs may be similar in patients with different biochemical defects.

Genetic Aspects

Families containing more than one individual with MM have been described, but the majority of cases are not familial. In a series of 71 patients, 18% of index cases had similarly affected relatives. No consistent pattern of inheritance was evident for any of the clinical syndromes or identified defects of mitochondrial metabolism, in either this study or other reports. Some pedigrees suggest autosomal recessive or dominant inheritance. No convincing X-linked pedigrees have been described. It is clear that when individuals are affected in more than one generation, maternal transmission to offspring is far more frequent than paternal transmission (in a ratio of approximately 9:1). Egger and Wilson suggested that this could be explained on the basis of mitochondrial inheritance. Support for this hypothesis comes from the fact that the majority of patients with MM have biochemical defects involving respiratory chain complexes containing subunits encoded by mitochondrial DNA.

Molecular Genetics of Mitochondrial Myopathies

Mitochondrial DNA Deletions   Initial restriction endonuclease analysis of mtDNA from the blood of patients with MM excluded major deletions of leukocyte mtDNA in patients, or any differences in restriction fragment patterns between normal and abnormal individuals in the same maternal line. Subsequently, Holt and colleagues showed that 9 of 25 patients with MM had two populations of mtDNA in muscle, one of which was deleted by up to 7 kb. None of these nine cases had detectably abnormal leukocyte mtDNA. The proportion of abnormal mtDNAs in muscle ranged from 20–70%. These observations have since been confirmed by others, and a close association between the presence of deletions and clinical involvement of the ocular muscles, sometimes with the other features of the

Kearns-Sayre syndrome, has been stressed. Tandem duplications of one population of mtDNAs have been described in two patients.

Thirty of 72 adult patients with MM had a deleted population of mtDNA in muscle. The deleted region was mapped in 25 cases; 13 had similar if not identical deletions extending over about 5 kb within the region 8,000–13,600 bp (Figure 12.2). None of the deletions definitely involved the origins of transcription or replication of either the heavy or light chains of mtDNA. All of the 30 patients with deletions of muscle mtDNA presented clinically with ophthalmoplegia and proximal myopathy, compared with 10 of 42 patients without detectable deletions, and 8 of these 30 patients had the Kearns-Sayre syndrome. None of the 11 patients presenting with proximal myopathy without ophthalmoplegia or the 19 with predominant CNS disease had deletions of muscle mtDNA.

Apart from the patients with the Kearns-Sayre syndrome, the patients with demonstrable deletions were relatively mildly affected clinically. This was reflected by biochemical studies of muscle mitochondria, which generally showed a mild reduction in respiratory capacity. Polarography was normal in 6 out of 18 cases investigated (Table 12.2). The site of the

Table 12.2: Polarographic Defects in Patients with MM and mtDNA Deletions

Defect Involving	Case Numbers (see Figure 12.2)
Complex I	2, 16*, 25*, 85*
Predominantly complex I	27[†], 83
Complexes I–III	1[†],43, 48[†]
Complexes I–IV	17[†], 26[†], 45[†]
Normal polarography	30[†], 44[†], 67[†], 68[†], 69[†], 70

See Holt et al. (1989) for definition of polarographic defects.
*Identical deletion, approximately 2.6 kb, between 11,854 and 14,471 bp.
[†]Identical deletion, 4.9 kb, between 8,470 and 13,459 bp.

Figure 12.2: Linearized map of the mitochondrial genome (boxed region), also showing (from top to bottom) extent of deleted regions in 25 patients with mitochondrial myopathy, and location of coding regions. The dotted lines at either end of the deletions represent their upper and lower limits, defined by the presence or absence of restriction sites. A = ATPase; CO = cytochrome oxidase; N/C = noncoding region; ND = NADH CoQ reductase subunits; O_H, O_L = origins of heavy and light strand replication; 12S and 16S = ribosomal RNAs; ⊢ = tRNAs. *Modified with permission from Holt IJ, Harding AE, Cooper JM, et al. Mitochondrial myopathies: clinical and biochemical features in 30 cases with major deletions of muscle mitochondrial DNA. Ann Neurol 1989;26:699–708.*

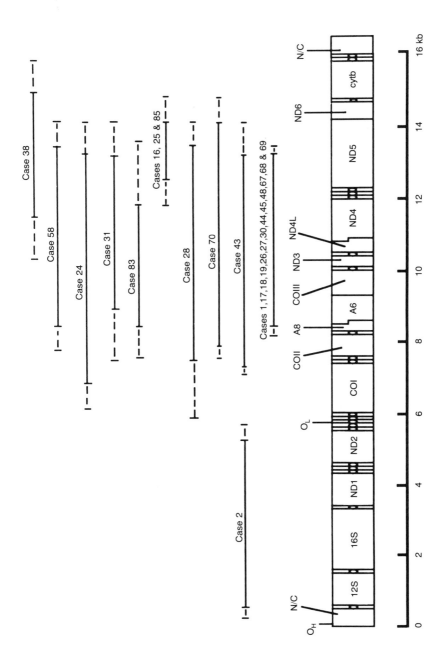

respiratory chain defect was variable, even in patients with seemingly identical deletions (Table 12.2; Figure 12.2). There was some correlation between the site of the biochemical defect and the deleted mitochondrial genes; in the four patients with pure complex I defects, the deletion either exclusively involved the genes for ND5 and 6, and tRNAs (patients 16, 25, 85), or the ND1 and 2 reading frames together with ribosomal RNA and tRNA genes (patient 2). This was in contrast to the 14 patients with more extensive loss of respiratory chain activity or normal biochemistry, in whom the deletions included reading frames for three or more subunits of complex I, one or two subunits of complex IV, and two subunits of complex V, as well as several tRNAs. It is of interest that the 10 patients with apparently identical deletions that were investigated biochemically had different biochemical and clinical features. Five of the six patients with normal muscle biochemistry were in this group.

There was no obvious correlation between the proportion of abnormal mtDNA in the muscle sampled and either clinical or biochemical severity. All the deletions involved a number of tRNAs. These would be expected to have a detrimental effect on translation of all mitochondrially encoded subunits of the respiratory chain, unless deleted mtDNAs and normal molecules are able to share tRNAs within individual mitochondria. The functional effects of deleted mtDNAs must depend on their distribution within mitochondria, muscle fibers, and different tissues. It is also possible that the total population of mtDNAs is increased in these patients; the relative proportion of deleted mtDNAs would have less functional significance in these circumstances. The different polarographic findings in patients with identical deletions may therefore reflect the absolute amount of normal, rather than deleted, mtDNA within individual mitochondria or its distribution in the whole mitochondrial population. The variable clinical features of these patients are presumably related to the amounts of normal and deleted mtDNA in muscle, brain, retinal pigment epithelium, and other tissues.

The survival of deleted mtDNA molecules in muscle is compatible with the observation that the number of muscle fibers does not increase significantly after early fetal life. The same may apply to CNS tissue in some cases of MM presenting with predominant CNS disease. A high proportion (72%) of deleted mtDNAs was observed in the brain of one patient with the Kearns-Sayre syndrome, and 52% of liver mtDNAs were also deleted. Frequent cell division in leukocyte precursors would be expected to select against cells containing genetically defective mitochondria. This is supported by studies of cultured muscle cells from patients with MM. Only one case of MM, with the Kearns-Sayre syndrome, has been shown to have a small proportion of deleted mtDNAs in fibroblasts and leukocytes by Southern blotting, although low-abundance deleted mtDNAs have been detected in patients' blood using the polymerase chain reaction.

The origin of muscle mtDNA deletions is unclear, in terms of both timing and molecular mechanism. Only two reported patients with deletions had clinically affected relatives. It seems probable that in most cases the deletions arose during oogenesis, with random partitioning of the two populations of mtDNA occurring during fetal development. Muscle mtDNA deletions were undetectable using Southern blotting in the mothers of three patients.

The deletion junction has been sequenced after amplification of the flanking region using the polymerase chain reaction in more than 50 patients to date, including 28 investigated by Mita and colleagues. In patients with the most common deletion (within the region 8,000–13,600 bp), sequences matching the published mtDNA sequence were observed up to 8,482 and down from 13,460 bp; these nucleotides were bridged by a 13-nucleotide (nt) direct repeat occurring at 8,470–8,482 and 13,447–13,459 bp of the published sequence. As this 13-nt sequence may be derived from either side of the flanking region, or be contributed in part by both sides, the exact breakpoint of the deletion could not be determined. Overall, approximately 50% of deletions (in 70% of patients) are flanked by perfect direct repeats 5–13 bp in length; the others are not. Possible mechanisms for deletion formation thus include recombination events mediated by enzymes recognizing short homologies. However, recombination is not thought to occur in mtDNA, partly because it does not seem to have orthodox DNA repair systems. Short direct repeats that flank sites susceptible to deletion have been observed in bacteria and the human beta globin genes. Such deletions are thought to arise as a result of slippage during replication, but this is difficult to envisage in mtDNA since its method of replication is different from that of nuclear DNA, using displacement, rather than discontinuous, synthesis.

Mitochondrial DNA Point Mutations and Other Genetic Defects in Mitochondrial Myopathies

Given that the excess of maternal inheritance provided a major impetus for investigating the mitochondrial genetic hypothesis in MM, it is perhaps slightly ironic that the first mtDNA defects identified, large deletions, are not usually inherited. A point mutation of mtDNA is more likely to be transmitted than mtDNAs with large deletions, as has been recently demonstrated in a large kindred with maternally inherited MERRF. This was an A-to-G transition at position 8,344 (a conserved nucleotide) in the lysine tRNA gene, which also occurred in two other unrelated patients with MERRF but none of 75 control subjects. All the patients had a mixture of mutant and normal mtDNA (heteroplasmy), the latter ranging from 2–27%. Disease severity showed some correlation with the proportion of mutant mtDNA when age was taken into account; for a given proportion of normal mtDNA, older subjects were more likely to have manifestations of MERRF than younger ones. The 8,344 mutation was also demonstrated in five Italian pedigrees with MERRF, but it was absent in two other cases of MERRF and patients with other MM phenotypes.

The phenotype associated with the MELAS mutation, also heteroplasmic and in a tRNA gene at base pair 3,423, is much more variable. Only about 50% of patients present with multiple stroke-like episodes, and others may have a combination of myopathy, ataxia and deafness, progressive external ophthalmoplegia, or even myopathy alone. Some patients with either the MERRF or MELAS mutations do not have ragged red fibers on muscle biopsy. Each mutation can be detected in leukocyte DNA from at least 95% of patients, providing a rapid screening test for mitochondrial disease. A maternally inherited syndrome of adult-onset myopathy and cardiomyopathy that segregated with a mtDNA mutation in the leucine tRNA gene has also been reported. There was a correlation between the amount of mutant mtDNA and both metabolic and clinical severity. Other possibly pathogenic point mutations of mtDNA, largely described in single kindreds, were reviewed by Wallace and colleagues in 1991.

It is unlikely that all cases of MM are due to defective mitochondrial genes, even on statistical grounds, as the nuclear genome codes for the majority of the respiratory chain subunits, as well as controlling their transport into mitochondria and subsequent assembly into functional enzyme complexes. Transcription and translation of the mitochondrial genome are also dependent on the nucleus. Evidence that MM may be caused by mutant nuclear genes is provided by Schapira and colleagues, who showed that some patients with complex I defects have a specific deficiency of the 24 kd FeS protein that is a nuclear product. It is also clear that nuclear genetic defects can cause mtDNA deletions in some instances. Zeviani and colleagues reported a family exhibiting autosomal dominant inheritance, with paternal transmission, in which affected members had multiple muscle mtDNA deletions of variable length, all originating in the D loop region. It was suggested that this disorder was caused by a nuclear defect involving mtDNA replication. A presumed autosomal recessive disorder giving rise to infantile lactic acidosis was associated with mtDNA depletion in affected tissues, including brain, liver, and kidney.

PEARSON'S SYNDROME

The common mtDNA deletion reported in patients with MM has been observed in lymphocytes from a patient with Pearson's syndrome, a combination of pancreatic, hepatic, and renal insufficiency with pancytopoenia. Many patients with Pearson's syndrome die in early childhood. We recently studied an 8-year-old boy who presented with Pearson's syndrome as a neonate, but survived to develop tremor, ataxia, ophthalmoplegia, and retinopathy at the age of 6 years. He thus had the features of both Pearson's syndrome and MM, and the latter was confirmed histologically. mtDNA analysis showed the common deletion in both leukocytes and muscle.

A POINT MUTATION OF mtDNA ASSOCIATED WITH A MULTISYSTEM NEUROLOGICAL DISEASE

A variable combination of developmental delay, retinitis pigmentosa, dementia, seizures, ataxia, proximal neurogenic muscle weakness, and sensory neuropathy was reported in four members of a pedigree exhibiting maternal inheritance. There was no histochemical evidence of mitochondrial myopathy on muscle biopsy. Blood and muscle from the patients contained two populations of mtDNA, one of which had a previously unreported restriction site for *Ava* I. Sequence analysis showed that this was due to a point mutation at nucleotide 8,993, resulting in an amino acid change from a highly conserved leucine to arginine in subunit 6 of mitochondrial H[22]-ATPase. There was some correlation between clinical severity and the amount of mutant mtDNA in the patients. Mutant mtDNA was present in only small quantities in the blood of healthy elderly relatives in the same maternal line.

LEBER'S HEREDITARY OPTIC NEUROPATHY

Leber's hereditary optic neuropathy (LHON) gives rise to acute or subacute bilateral visual loss, usually in young adult males. In the acute phase the optic discs are swollen with tortuous retinal arterioles, and telangiectases are present in the peripapillary small vessels. There is evidence that the appearance of capillary microangiopathy predates the onset of symptoms. The visual field loss consists initially of an enlarged blind spot; this increases to involve central vision, producing a large centrocecal scotoma. Optic atrophy is apparent within two months. Loss of visual acuity is generally severe (6/60 or less), although limited improvement is sometimes observed. There is a curious association between LHON and cardiac pre-excitation syndromes.

About 85% of patients with LHON are male, and 18% of female carriers are affected. Paternal transmission of the disease to daughters or grandchildren has never been described, making X-linked inheritance un-likely. Between 70–100% of daughters of female carriers are also carriers, and 50–100% of the sons of carriers are affected. This pattern of transmis-sion suggests cytoplasmic or mitochondrial inheritance. On the basis of these observations, and also because of the finding of enlarged mitochon-dria with proliferation of cristae in skeletal muscle from patients, Nikoske-lainen and colleagues suggested that LHON could be caused by a mutation of mtDNA.

Wallace and colleagues reported a point mutation at position 11,778 of mtDNA in members of 9 out of 11 LHON pedigrees; this was not present in 45 control subjects. The mutation led to an amino acid change from arginine to histidine in a subunit of NADH CoQ reductase (ND4). The mutation was present in all maternally related individuals in these families, regardless of whether or not they were affected. It also appeared to be homoplasmic, that is, present in all mtDNA molecules analyzed. The change from adenine to guanine at position 11,778 in mtDNA results in loss of a restriction site for the restriction endonuclease *Sfa*NI. Using this observation, Holt and colleagues investigated eight LHON families in the United Kingdom. The 11,778 bp mutation was present in only four kin-dreds; all of the other four contained members with clinically typical LHON, which was maternally transmitted. There was thus evidence of genetic heterogeneity, and this had some clinical correlates. Recovery of useful visual function was not observed in members of any of the four families with the 11,778 bp mutation. Some improvement of vision had occurred, after periods ranging between one and four years, in members of all four of the families in which the *Sfa*NI site was intact.

In the majority of individuals in the pedigrees with the 11,778 bp mutation, Holt and colleagues showed that there was a mixture of mutant and normal mtDNA in peripheral blood (Figure 12.3). This was not always detected after short periods of autoradiography, but became obvi-ous after two to seven days' exposure in all but one affected male. All affected males and females had at least 96% mutant mtDNA, and the same applied to unaffected women with affected sons. Less than 80% mutant mtDNA was observed in some unaffected males at risk, and in a female with healthy sons. Carrier females with a relatively high proportion of normal mtDNA should be less likely to transmit the disease to their sons if these findings in peripheral blood reflect the populations of mtDNA in

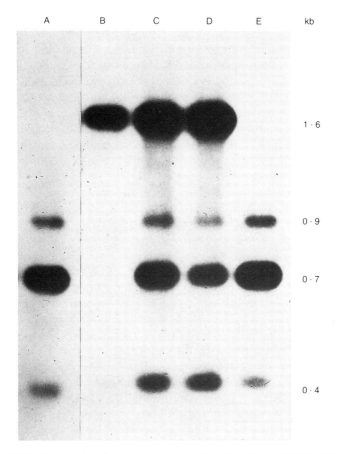

Figure 12.3: Restriction fragment patterns after digestion of DNA with *Sfa*NI and hybridization to a fragment of HeLa cell mtDNA (11,680–12,570 bp). Lane A: normal control shows fragments 0.9, 0.7, and 0.4 kb in length. Lane B: in a male with LHON, the 0.7 kb and 0.9 kb bands are replaced by one of 1.6 kb due to loss of the *Sfa*NI site at 11,778 bp (0.4 kb band not visible because of unequal DNA loading). In lanes C and D (one unaffected at-risk male and one possible carrier female), there is a mixture of normal and mutant mtDNA. In lane E, another unrelated male with LHON, the *Sfa*NI site is retained. The sample in lane B showed evidence of heteroplasmy after longer exposure. *Reprinted by permission from Holt IJ, Miller DH, Harding AE. Genetic heterogeneity and mitochondrial DNA heteroplasmy in Leber's hereditary optic neuropathy. J Med Genet 1989;26:739–743.*

ova. Similarly, the proportion of mutant mtDNA in the optic nerves or their vasculature could determine the likelihood of developing the disease. Subsequently, three further mtDNA mutations have been described that are exclusively found in affected families. The second most frequent is at position 3,460 bp in the *ND1* subunit gene. The position 4,160 bp mutation (also in the *ND1* gene) has been reported exclusively in a large unusual Australian pedigree referred to below. The last is at 5,244 bp in the *ND2* gene. Other mtDNA mutations have been reported in LHON pedigrees, but their significance is unclear since they also occur in the normal population. Detection of these mutations is useful in clinical practice and can be used as a confirmatory test, particularly in patients without affected relatives.

It is not clear how an amino acid change in a subunit of complex I of the respiratory chain causes the development of subacute blindness in young adults. The selective nature of the disease is particularly difficult to explain since the LHON mtDNA mutation, although heteroplasmic, is present in a high proportion of leukocyte mtDNAs. It is of interest that reduced NADH CoQ reductase (complex I) activity in platelets was reported in an unusual Australian family, some members of which had typical LHON. Other family members had clinical features suggesting a diagnosis of mitochondrial encephalomyopathy, including stroke-like episodes, a syndrome described in association with complex I defects. The preponderance of affected males in LHON families is compatible with, but not explained by, an mtDNA mutation. It is possible that environmental factors such as cigarette smoking are relevant in this context.

The presence of two populations of mtDNA (heteroplasmy), one normal and one mutant, in association with three human diseases (mitochondrial myopathies, LHON, and the 8,993 bp point mutation described earlier) is of interest. MtDNA heteroplasmy has never been described in control subjects, despite the high mutation rate of mtDNA, which implies a need for heteroplasmy during the transition from one genotype to another. Hauswirth and Laipis demonstrated heteroplasmy in a single maternal line of Holstein cows. They suggested that mtDNA could switch completely from one genotype to another in a single generation if the number of mtDNAs is greatly reduced at some point in oogenesis, as probably occurs in maturing ova. These observations suggest that persistent heteroplasmy and deleterious mtDNA mutations are related in some way. It is possible that the rapid switch from one mtDNA type to another, which seems to occur during the evolution of harmless, and possibly advantageous, polymorphisms, does not take place when mtDNA mutations are harmful, because of selection in favor of normal mtDNAs. Survival would be less likely if these mutations were homoplasmic.

OTHER DISEASES POSSIBLY ASSOCIATED WITH DEFECTS OF THE MITOCHONDRIAL GENOME

To date, mtDNA defects have been described in association with a wide range of clinical syndromes, and there are other possible candidates for mitochondrial genetic disease. One very speculative example is Parkinson's disease (PD). A selective deficiency of complex I has been reported in the substantia nigra and platelets of PD patients. Although PD is often familial, a substantial genetic etiological component has largely been discounted on the basis of similar concordance rates for the prevalence of the disorder in dizygotic and monozygotic twins. A heteroplasmic mtDNA defect is compatible with this observation.

SELECTED REFERENCES

Attardi G. The elucidation of the human mitochondrial genome: a historical perspective. BioEssays 1986;5:34–39.

Cann RL, Stoneking M, Wilson AC. Mitochondrial DNA and human evolution. Nature 1987;325:31–36.

DiMauro S, Bonilla E, Zeviani M, Nakagawa M, DeVivo DC. Mitochondrial myopathies. Ann Neurol 1985;17:521–538.

Egger J, Wilson J. Mitochondrial inheritance in a mitochondrially mediated disease. N Engl J Med 1983;309:142–146.

Goto Y, Nonaka I, Horai S. A mutation in the tRNA$^{Leu(UUR)}$ gene associated with the MELAS subgroup of mitochondrial encephalomyopathies. Nature 1990; 348:651–653.

Hammans SR, Sweeney MG, Brockington M, et al. Mitochondrial encephalopathies: molecular genetic diagnosis from blood samples. Lancet 1991;337: 1311–1313.

Harding AE, Petty RKH, Morgan-Hughes JA. Mitochondrial myopathy: a genetic study of 71 cases. J Med Genet 1988;25:528–535.

Hatefi Y. The mitochondrial electron transport and oxidative phosphorylation system. Ann Rev Biochem 1985;54:1015–1069.

Hauswirth WW, Laipis PJ. Transmission genetics of mammalian mitochondria: a molecular model and experimental evidence. In: Quagliariello E, Slater EC, Palmieri F, Saccone C, Kroon AM, eds. Achievements and perspectives of mitochondrial research, volume II, biogenesis. Elsevier: Amsterdam, 1985:49–60.

Holt IJ, Harding AE, Morgan-Hughes JA. Deletions of mitochondrial DNA in patients with mitochondrial myopathies. Nature 1988;331:717–719.

Holt IJ, Miller DH, Harding AE. Genetic heterogeneity and mitochondrial DNA heteroplasmy in Leber's hereditary optic neuropathy. J Med Genet 1989; 26:739–743.

Holt IJ, Harding AE, Petty RKH, Morgan-Hughes JA. A new mitochondrial disease associated with mitochondrial DNA heteroplasmy. Am J Hum Genet 1990;46:428–433.

Holt IJ, Harding AE, Cooper JM, et al. Mitochondrial myopathies: clinical and biochemical features in 30 cases with major deletions of muscle mitochondrial DNA. Ann Neurol 1989;26:699–708.

Howell N, Kubacka M, Xu M, McCullough DA. Leber hereditary optic neuropathy: involvement of the mitochondrial ND1 gene and evidence for an intragenic suppressor mutation. Am J Hum Genet 1991;48:935–942.

Huoponen K, Vilkki J, Aula P, et al. A new mtDNA mutation associated with Leber hereditary optic neuroretinopathy. Am J Hum Genet 1991;48:1147–1153.

McShane MA, Hammans SR, Sweeney M, Holt IJ, Beattie TJ, Brett EM, Harding AE. Pearson's syndrome and mitochondrial encephalomyopathy in a patient with a deletion of mitochondrial DNA. Am J Hum Genet 1991;48:39–42.

Mita S, Rizzuto R, Moraes CT, et al. Recombination via flanking direct repeats is a major cause of large-scale deletions of human mitochondrial DNA. Nucleic Acids Res 1990;18:561–567.

Moraes C, DiMauro S, Zeviani M, et al. Mitochondrial DNA deletions in progressive external opthalmoplegia and Kearns-Sayre syndrome. N Engl J Med 1989;320:1293–1299.

Moraes CT, Schon EA, DiMauro S, Miranda AF. Heteroplasmy of mitochondrial genomes in clonal cultures from patients with Kearns-Sayre syndrome. Biochem Biophys Res Commun 1989;160:765–771.

Moraes CT, Shanske S, Tritschler H-J, et al. mtDNA depletion with variable tissue expression: a novel genetic abnormality in mitochondrial diseases. Am J Hum Genet 1991;48:492–501.

Nikoskelainen E. New aspects of the genetic, etiologic and clinical puzzle of Leber's disease. Neurology 1984;34:1482–1484.

Parker WD, Boyson SJ, Parks JK. Abnormalities of the electron transport chain in idiopathic Parkinson's disease. Ann Neurol 1989;26:719–723.

Parker WD, Oley CA, Parks JK. Deficient NADH: coenzyme Q oxidoreductase activity in Leber's hereditary optic neuropathy. N Engl J Med 1989;320:1331–1333.

Petty RKH, Harding AE, Morgan-Hughes JA. The clinical features of mitochondrial myopathy. Brain 1986;109:915–938.

Poulton J, Deadman ME, Gardiner RM. Duplications of mitochondrial DNA in mitochondrial myopathy. Lancet 1989;1:236–240.

Rosing HS, Hopkins LC, Wallace DC, et al. Maternally inherited mitochondrial myopathy and myoclonic epilepsy. Ann Neurol 1985;17:228–237.

Rotig A, Colonna M, Bonnefont JP, et al. Mitochondrial DNA deletion in Pearson's marrow/pancreas syndrome. Lancet 1989;1:902–903.

Schapira AHV, Cooper JM, Dexter D, Clark JB, Jenner P, Marsden CD. Mitochondrial complex I deficiency in Parkinson's disease. J Neurochem 1990;54;823–827.

Shoffner JM, Lott MT, Lezza AMS, Seibel P, Ballinger SW, Wallace DC.

Myoclonic epilepsy and ragged-red fiber disease (MERRF) is associated with a mitochondrial DNA tRNA^lys mutation. Cell 1990;61:931–937.

Wallace DC, Lott MT, Torroni A, Shoffner JM. Report of the committee on human mitochondrial DNA. Cytogenet Cell Genet 1991;58:1103–1123.

Wallace DC, Singh G, Lott MT, et al. Mitochondrial DNA mutation associated with Leber's hereditary optic neuropathy. Science 1988;242:1427–1430.

Zeviani M, Gellera C, Antozzi C, et al. Maternally inherited myopathy and cardiomyopathy: association with a new mutation in the mitochondrial DNA tRNA^Leu(UUR). Lancet 1991;338:143–147.

Zeviani M, Servidei S, Gellera C, Bertini E, DiMauro S, DiDonato S. An autosomal dominant disorder with multiple deletions of mitochondrial DNA starting at the D-loop region. Nature 1989;339:309–311.

Zeviani M, Amati P, Bresolin N, et al. Rapid detection of the A→G^(8344) mutation of mtDNA in Italian families with myoclonus-epilepsy and ragged-red fibers (MERRF). Am J Hum Genet 1991;48:203–211.

CHAPTER 13

Fragile X Syndrome

David L. Nelson

Fragile X (Martin–Bell) syndrome is variable in its neurological phenotype, ranging from mild learning disability to profound mental retardation. The disorder is second only to Down syndrome as a genetic cause of mental retardation, and is the most common form of familial mental retardation, with a frequency in boys estimated as high as 1 in 1,250. Fragile X syndrome segregates as an X-linked dominant with reduced penetrance since either sex, when carrying the fragile X mutation, may exhibit mental deficiency. Sherman et al. have shown that approximately 30% of carrier females are penetrant and that 20% of males carrying the fragile X chromosome are phenotypically normal but may transmit the disorder and have fully penetrant grandsons. In addition to the mental retardation, which is variable in severity, penetrant males exhibit additional phenotypic involvement, including macroorchidism and distinctive facies. Since fully penetrant males rarely reproduce, it has been suggested that the frequency of new fragile X mutations may be as high as 1 in 3,000 germ cells to maintain the population frequency.

Fragile X syndrome, as implied by the name, is associated with a fragile site, expressed as an isochomatid gap in the metaphase chromosome, at map position Xq27.3. The fragile X site is induced by cell culture conditions that perturb deoxypyrimidine pools and is rarely observed in greater than 50% of the metaphase spreads. Until recently, neither the molecular nature of the fragile X site nor its relationship to the gene(s) responsible for the clinical expression of the syndrome were understood. However, based upon genetic linkage studies as well as in situ hybridization, the fragile X site and its associated gene(s) are tightly linked, if not coincident.

Fragile X syndrome is difficult to ascertain clinically prior to age two.

At this point, developmental delay and somatic features such as increased head circumference and prominence of the ears, forehead, and jaw become more prevalent. In males at puberty, macroorchidism is common. Mental retardation is variable, but over 90% of male patients have IQ scores in the 20–60 range, with a mean between 30 and 45. In the 30% of females who are affected, the degree of mental deficiency ranges from mild learning disorders to severe retardation. Some overlap with features of autism has been described, and a significant percentage of males with autism as a primary diagnosis exhibit a fragile site in Xq27.3. No significant pathologic features have been described to account for the mental deficiency found in fragile X syndrome.

GENETIC ASPECTS

Inheritance of the fragile X syndrome is quite complicated, although the features of the mutation at the DNA level (see below) are beginning to provide an explanation for the unusual genetics found in this disorder. The most striking aspect of fragile X is its incomplete penetrance in both males and females with the mutation. This is particularly interesting in the case of "normal transmitting males" (NTMs), who transmit the mutation to grandsons but are unaffected themselves. More complicated are the probabilities of mental deficiency based on the affected status of parents. These are summarized in Figure 13.1, and demonstrate the peculiar nature of inheritance in this disease. In brief, the probability of mental retardation is increased by the number of generations through which the mutation has passed, and is higher for both sons and daughters of affected females.

To account for these data, models have been proposed for the nature of the mutation in fragile X. Nussbaum and Ledbetter proposed the presence of a repetitive sequence that is modified once as a premutation, then grows in subsequent generations (especially in the daughters of NTMs) to a longer repeat that confers disease. This would account for transmitting males (the premutation not having a phenotypic effect), and for the higher probability of affected offspring from daughters of transmitting males than from the mothers of transmitting males (see Figure 13.1). Recent work in several groups demonstrates the presence of a simple tandem repeat at the fragile site that is variable in length, and larger in affected individuals.

The number of copies of the CGG repeat in fragile X premutations varies from 52 to over 200 by the PCR assay. One remarkable aspect of these alleles is their complete instability in meiotic transmission. Alleles in children always differ from those in the parent; these changes can involve alterations within the premutation range or expansions to the full mutation. Risk of expansion to the full mutation increases with the size of the premu-

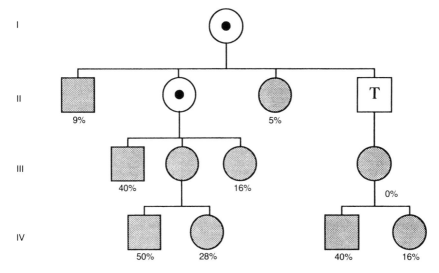

Figure 13.1: Artificial pedigree of fragile X syndrome family. A female carrier of normal intelligence is symbolized by an outlined circle with a central dot. A transmitting male is symbolized by a square containing a T. The risk for mental retardation for a particular individual is shown as a percent figure. *Adapted from Nussbaum RL, Ledbetter DL. Fragile X syndrome: a unique mutation in man. Ann Rev Genet 1986;20:109–145.*

tation, and this observation provides a molecular explanation of the variation in risk of mental retardation with pedigree position (see Figure 13.1). In general, the size of the premutation increases in each generation, and this fact, coupled with larger risks of expansion to the full mutation in female meiosis, accounts for the increasing risk of mental retardation in subsequent generations. With the identification of a different unstable triplet repeat in the gene that is defective in myotonic dystrophy, the role of unstable DNA in deviations from Mendelian expectations in human genetic disease (penetrance in fragile X, expressivity and anticipation in myotonic dystrophy) has been established. Additional genes and phenomena await exploration.

An alternative hypothesis was formulated by Laird, in which the local region including the fragile site, if mutant, frequently fails to reactivate once inactivated. This would account for the presence of transmitting males, since their mutations would have to be present on an inactive X chromosome in their daughters, fail to reactivate in oogenesis, and be transmitted to grandsons and granddaughters in an inactive form. Thus, passage through the female germ line would be requisite to the development of an affected phenotype, and could involve other modifications peculiar to this passage, such as methylation. The recent demonstration of methylation of a CpG island at the fragile X site that is highly (but not completely) specific to affected fragile X patients provides some evidence that this mechanism may be relevant.

DIAGNOSIS

Diagnosis of fragile X has been performed by cytogenetic expression of the fragile site in Xq27.3. This has been a fairly accurate test in proficient cytogenetics laboratories and is effective at identifying affected males. In females, it has been much more problematic, and cytogenetic expression is not a good indicator of affected status nor is it an ideal test for carriers. Linkage analysis has been used extensively as a diagnostic tool; however, the polymorphic markers that have been used are significantly distant from the gene so that the risk of recombination in the chromosome of interest is reasonably high. Recently, a CA repeat polymorphism that is highly informative and exhibits no recombination with the disorder was identified ~100 kb proximal to the fragile site. The colocalization of a genetic marker with no recombination and the fragile site is significant, as it demonstrates that the site and the syndrome are linked.

APPROACHES TO ISOLATION OF THE FRAGILE SITE

In recent years, several laboratories invested significant efforts in developing approaches to isolation of the fragile site in Xq27.3. Numerous random DNA markers were identified; those capable of recognizing polymorphisms were used to type families to determine linkage distances, and panels of somatic cell hybrids with X chromosome translocations and deletions were also used to develop physical maps of cloned fragments without polymorphism. Difficulties with generation of DNA fragments from the region led to more exotic approaches, such as microdissection of the fragile-site region and PCR amplification of fragments from the region. Physical maps based on long-range restriction mapping were also developed, and these provided the initial indication of fragile X–specific methylation at a CpG island.

A series of somatic cell hybrids with X chromosome translocations at or very close to the fragile site was developed by the Warren laboratory. This panel was instrumental in identification of the fragile site and was used by all groups involved in the search. The observation that a fragile X chromosome retained in a hamster cell background maintained its fragile character allowed production of translocations at the fragile site by induction of the site followed by selection for and against markers (*HPRT* and

G6PD) flanking the fragile site. Numerous hybrid cell lines containing either Xpter-Xq27.3 or Xq27.3-Xqter were developed. The finding that many of the translocation chromosomes retained the ability to induce the fragile site indicated that at least a portion of the site had been retained on the broken X chromosome. As DNA markers were found close to the region, this panel of hybrids allowed determination of the location relative to the fragile site, and in some rare cases, fine discrimination of the position of the probe. The breakpoints in these chromosomes were used by all groups to identify the fragile site (see below), and retrospective analysis has shown that the majority of the breakpoints occurred at the fragile X site, within a few kilobase pairs of DNA.

MOLECULAR BASIS

The early 1990s have seen the identification of very closely linked DNA fragments combined with the analysis of the somatic cell hybrids with breakpoints at the fragile site, the description of the region, and the isolation of large insert clones to characterize the site. Markers closely linked and flanking the hybrid breakpoints were identified. Pulsed field gel electrophoresis was used in conjunction with close markers to determine that a CpG island was preferentially methylated in fragile X patients, but not in normal males. Yeast artificial chromosome (YAC) clones that spanned the fragile site as defined by the somatic cell hybrids and contained the CpG island were reported (Figure 13.2).

Narrowing the region of breakpoints in the somatic cell hybrids and closing in on the affected CpG island turned out in several groups to identify the fragile site at the DNA level. A 5.2-kb *Eco*RI fragment contains the majority of hybrid breakpoints, the CpG island, an exon of the *FMR1* gene (see below), and an unusual repetitive sequence of $(CGG)_n$ that demonstrates variation in normal individuals, with significant increases in length in affected fragile X patients. All these features can be further confined to a 1-kb *Pst* I fragment (Figure 13.3).

Family studies confirm that the increase in length in this region is the basis for fragile X syndrome, and that the CGG repeats are responsible for the increasd length of the DNA fragments. Normal transmitting males demonstrate a premutation, with a small increase in length that is not within the range of normal variation, but is apparently insufficient to confer disease. In daughters of NTMs, the premutation is maintained, but grandchildren exhibit significant size increases at the repeat locus. The instability of the repeat when passed through the female germ line appears to account for the increased probability of mental retardation in successive generations of the pedigree. The high mutation frequency of this fragile X

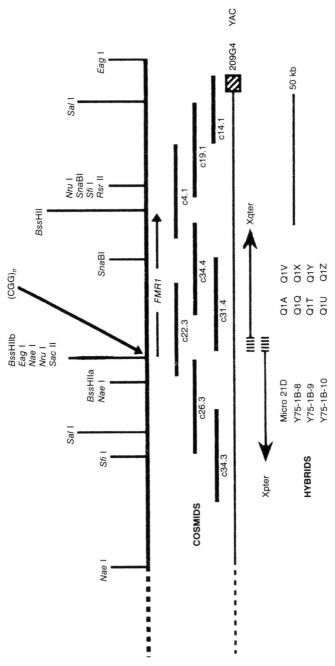

Figure 13.2: Map of the fragile X site in Xq27.3. Shown is ~200 kb of DNA surrounding the fragile X site in Xq27.3. A rare-cutter restriction map derived from the YAC 209G4, the positions of cosmid subclones from the YAC, and the locations of the breakpoints in proximal- and distal-retaining somatic cell hybrid breakpoints are indicated. The position of the (CGG)$_n$ repeats adjacent to the CpG island containing BssHII site b is indicated with an arrow. *Reprinted by permission from Verkerk AJMH, Pieretti M, Sutcliffe JS, et al. Identification of a gene (FMR-1) containing a CGG repeat coincident with a breakpoint cluster region exhibiting length variation in fragile X syndrome. Cell 1991;65:905–914.*

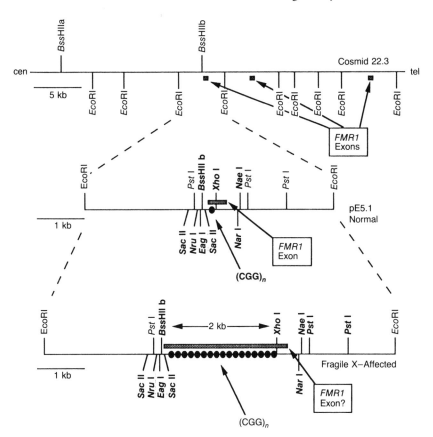

Figure 13.3: Map of the fragile X site in cosmid 22.3, subclone pE5.1, and the presumed structure in an affected fragile X chromosome. Restriction maps and other relevant features are given for cosmid and plasmid subclones from YAC 209G4. The hypothesized structure of the fragile X mutation is shown as multiples of the CGG-repeat region represented by black circles. Rare-cutting restriction sites are in bold type.

DNA fragment is unprecedented. The mutant form is also found to be significantly unstable, as it is difficult to maintain in yeast or bacterial cloning systems, and its length is found to vary from cell to cell within a single individual.

The number of copies of CGG repeats in normal individuals is highly variable; alleles from 6 to 46 copies have been observed thus far; however, the majority of chromosomes contain 29 copies. Alleles in this size range appear to be stable in meiosis. Approximately 65% of females are found to be heterozygous. A PCR-based assay for the length of the CGG repeat has been developed and should provide a simple means for ruling out fragile X syndrome in suspected males.

The *FMR1* gene expresses an mRNA of 4.4 kb that is found in a wide variety of tissues. Clones totaling 3.8 kb of cDNA from this mRNA have been isolated from a human fetal brain library. These clones have identified a single long open reading frame extending from the 5′ end for 1,971 bp that predicts 657 amino acids of the *FMR1*-encoded polypeptide. Since no poly(A) sequence is found at the 3′ end, and since the 5′ end has not yet been identified (the open reading frame remains open), the location of the missing 600 bp of cDNA is unknown. Thus, it is unclear whether the entire coding sequence has been determined. If the 5′ end of the cDNA represents the 5′ end of the gene, then the methylation-sensitive CpG island is at the promoter of the gene.

Comparison of the nucleic acid and predicted peptide sequences for the *FMR1* gene to known sequences in data bases has not shed much light on the potential function of *FMR1*. The gene appears to have been highly conserved in evolution; it demonstrates related sequences in species as distant from human as *C. elegans* and yeast.

The presence of CGG repeats in the *FMR1*-coding region close to the 5′ end of the cDNA clones suggests that this repeat encodes a poly-arginine repeat in the *FMR1* protein. The finding that the CGG repeats responsible for the genomic DNA rearrangements in fragile X syndrome compose a portion of the coding sequence for an associated gene suggests that alterations in the repeat number could affect the protein or its expression. In most cases, levels of mRNA from the *FMR1* gene are greatly reduced in lymphoblastoid cell lines and fresh blood leukocytes derived from fragile X patients. However, some patients (~20%) exhibit near-normal levels of expression. This fraction of patients also demonstrates mosaicism for methylation at the CpG island, indicating that the *FMR1* gene is a highly likely candidate for the primary cause of the phenotype of fragile X syndrome. Determination of the normal role of *FMR1* and the consequences of its lack of expression in fragile X syndrome will likely lead to much greater insight into the disease.

CONCLUSIONS

Detailed investigation of the nature of fragile X syndrome can now commence. The first goals of identification of the fragile site and associated genes have been achieved. The tools provided by these achievements now allow detailed description of the nature of the mutation, and will uncover the mechanism of the peculiar genetic alterations found in this disease. These reagents also offer the potential for unambiguous diagnosis of the mutation and disease, and may eliminate the expensive and often inaccurate cytogenetic tests used currently. Isolation of the *FMR1* gene should allow insight into the generation of the phenotypic features of fragile X syndrome, in particular the mechanism of mental retardation. These insights may also provide a rational basis for designing treatments for the disease. The success of the gene hunt in fragile X has answered some of the questions about this strange genetic disorder; however, it has raised many more questions than it has answered, and much more effort will be required to fully understand this syndrome.

ACKNOWLEDGMENTS

I wish to acknowledge my many collaborators and colleagues in the fragile X community, in particular, Steve Warren, Ben Oostra, and Tom Caskey, with whom I have shared the excitement of this pursuit. The hard work and enthusiasm of all contributors, but especially Maura Pieretti, Ying-hui Fu, Annemieke Verkerk, and Jim Sutcliffe are also greatly appreciated. This work was supported by a grant from the U.S. Department of Energy (FG05-88ER60692).

SELECTED REFERENCES

Bell MV, Hirst MC, Nakahori Y, et al. Physical mapping across the fragile X: hypermethylation and clinical expression of the fragile X syndrome. Cell 1991;64: 861–866.

Brook JD, McCurrach ME, Harley HG, et al. Molecular basis of myotonic dystrophy: expansion of a trinucleotide (CTG) repeat at the 3′ end of a transcript encoding a protein kinase family member. Cell 1992;68:799–808.

Brown WT. The fragile X: progress towards solving the puzzle. Am J Hum Genet 1990;47:175–180.

Brown WT, Jenkins ED, Cohen IL, et al. Fragile X and autism: a multicenter survey. Am J Med Genet 1986;23:341–358.

Dietrich A, Kioschis P, Monaco AP, et al. Molecular cloning and analysis of the fragile X region in man. Nucl Acids Res 1991;19:2567–2572.

Fu Y-H, Kuhl DPA, Pizzuti A, et al. Variation of the CGG repeat at the fragile X site results in genetic instability: resolution of the Sherman paradox. Cell 1991;67:1047–1058.

Fu Y-H, Pizzuti A, Fenwick RG Jr, et al. An unstable triplet repeat in a gene related to myotonic muscular dystrophy. Science 1992;255:1256–1258.

Gustavson KH, Blomquist H, Holmgren G. Prevalence of fragile-X syndrome in mentally retarded children in a Swedish county. Am J Hum Genet 1986;23:581–588.

Hagerman RJ, Jackson AW, Levitas A, Rimland B, Braden M. An analysis of autism in fifty males with the fragile X syndrome. Am J Med Genet 1986;23:359.

Heitz D, Rousseau F, Devys D, et al. Isolation of sequences that span the fragile X and identification of a fragile X-related CpG island. Science 1991;251: 1236–1239.

Hirst M, Roch A, Flint TJ, et al. Linear order of new and established markers around the fragile site at Xq27.3. Genomics 1991;10:243–249.

Krawczun MS, Jenkins EC, Brown WT. Analysis of the fragile-X chromosome: localization and detection of the fragile site in high resolution preparations. Hum Genet 1985;69:209–211.

Laird CD. Proposed mechanism of inheritance and expression of the human fragile-X syndrome of mental retardation. Genetics 1987;117:587–599.

MacKinnon RN, Hirst MC, Bell MV, et al. Microdissection of the fragile X region. Am J Hum Genet 1990:47:181–187.

Mahadevan M, Tsilfidis C, Sabourin L, et al. Myotonic dystrophy mutation: an unstable CTG repeat in the 3′ untranslated region of the gene. Science 1992;255:1253–1255.

Nelson DL, Ballabio A, Victoria MF, et al. Alu PCR for regional assignment of 110 yeast artificial chromosome clones from the human X chromosome: identification of clones associated with a disease locus. Proc Natl Acad Sci USA 1991;88:6157–6161.

Nielsen KB. Diagnosis of the fragile X syndrome (Martin-Bell syndrome): clinical findings in 27 males with the fragile site at Xq28. J Ment Defic Res 1983;27:211.

Nussbaum RL, Ledbetter DL. The fragile X syndrome. In: Scriver CR, Beaudet AL, Sly WS, Valle D, eds. The metabolic basis of inherited disease. New York: McGraw Hill, 1990:327–341.

Nussbaum RL, Ledbetter DL. Fragile X syndrome: a unique mutation in man. Ann Rev Genet 1986;20:109–145.

Oberle I, Rousseu F, Heitz D, et al. Instability of a 550-base pair DNA segment and abnormal methylation in fragile X syndrome. Science 1991;252:1097–1102.

Oostra BA, Hupkes PE, Perdon LF, et al. New polymorphic DNA marker close to the fragile site FRAXA. Genomics 1990;6:129–132.

Pergolizzi RG, Erster SH, Goonewardena P, Brown WT. Detection of full fragile X mutation. Lancet 1992;339:271–272.

Pieretti M, Zhang F, Fu Y-H. et al. Absence of expression of the FMR-1 gene in fragile X syndrome. Cell 1991;66:817–822.

Rousseau F, Heitz D, Biancalana V, et al. Direct diagnosis by DNA analysis of the fragile X syndrome of mental retardation. N Engl J Med 1991;325:1673–1681.

Rousseau F, Vincent A, Rivella S, et al. Four chromosomal breakpoints and four new probes mark out a 10 cM region encompassing the FRAXA locus. Am J Hum Genet 1991;48:108–116.

Sherman SL, Jacobs PA, Morton NE, et al. Further segregation analysis of the fragile X syndrome with special reference to transmitting males. Hum Genet 1985;69:3289–3299.

Sherman SL, Morton NE, Jacobs PA, Turner G. The marker (X) syndrome: a cytogenetic and genetic analysis. Ann Hum Genet 1984;48:21–37.

Sutherland GR, Haan EA, Kremer E, et al. Hereditary unstable DNA: a new explanation for some old genetic questions? Lancet 1991;338:289–292.

Sutherland GR, Hecht F. Fragile sites on human chromosomes. New York: Oxford University Press, 1985:53.

Suthers GK, Callen DF, Hyland VJ, et al. A new DNA marker tightly linked to the fragile X locus (FRAXA). Science 1990;246:1298–1300.

Suthers GK, Hyland VJ, Callen DF, et al. Physical mapping of new DNA probes near the fragile X mutation (FRAXA) by using a panel of cell lines. Am J Hum Genet 1990;47:187–195.

Verkerk AJMH, Pieretti M, Sutcliffe JS, et al. Identification of a gene (FMR-1) containing a CGG repeat coincident with a breakpoint cluster region exhibiting length variation in fragile X syndrome. Cell 1991;65:905–914.

Vincent A, Heitz D, Petit C, Kretz C, Oberle I, Mandel J-L. Abnormal pattern detected in fragile X patients by pulsed-field gel electrophoresis. Nature 1991;349:624–626.

Warren ST, Knight SJL, Peters JF, Stayton CL, Consalez GG, Zhang F. Isolation of the human chromosomal band Xq28 within somatic cell hybrids by fragile X site breakage. Proc Natl Acad Sci USA 1990;87:3856–3860.

Warren ST, Zhang F, Licameli GR, Peters JF. The fragile X site in somatic cell hybrids: an approach for molecular cloning of fragile sites. Science 1987;237:420–423.

Webb TP, Bundey SE, Thake AI, Todd J. Population incidence and segregation ratios in the Martin-Bell syndrome. Am J Hum Genet 1986;23:573–580.

Yu S, Pritchard M, Kremer E, et al. The fragile X genotype is characterized by an unstable region of DNA. Science 1991;252:1179–1181.

Index

Acceptor sites, internal, 14*f*, 15
α-Actinin, 81, 84
Activation, gene, 8
Agarose gel electrophoresis. *See* Gel electrophoresis
Alleles, 3
   haplotype analysis, 36, 38
   RFLP, 21*f*, 22*f*, 23
Allele-specific oligonucleotide (ASO) hybridization, 23
   PCR generation, 36, 37*f*
   phenylalanine hydroxylase, 188, 193, 194*f*
Allopurinol, 141, 142
Alternative 3′- and 5′-terminal exons, 16
Alternative splicing. *See* Splicing, alternative
Alu polymerase chain reaction, 32–33
Alu sequences, 5, 6
   Duchenne muscular dystrophy gene, 95
   HPRT gene, 132, 132*f*, 133, 134*f*
   neurofibromatosis 1 gene, 173
Alzheimer's disease, 56, 67
Amino acids, 4
Amino acid sequences
   degeneracy of genetic code, 32
   transthyretin, 62–63*f*
Amino acid substitutions. *See* Point mutations
Amyloid A, serum (SAA), 57, 57*t*
Amyloidosis, hereditary
   classification, 55–56, 57*t*
   mendelian heritability, 59–61, 60*f*

molecular medicine, 67–70
   diagnostics, 67–69, 68*t*, 69*f*
   therapeutic options, 67–70
mutation sites, 59–61, 60*f*
   apolipoprotein A-1 form, 66
   localized forms, 67
   transthyretin molecule, 61–66, 62*f*, 63*f*, 64*f*, 65*t*
pathogenesis, 56–58, 58*f*, 59*f*
physical chemical characteristics, 56, 57*t*
Amyloid P, serum (SAP), 56
Angiofibromas. *See* Tuberous sclerosis complex
Animal models. *See also* Mouse models
   Charcot-Marie-Tooth disease, 241
   Duchenne muscular dystrophy, 104–105
Antibodies. *See* Immunochemistry
Antiparallel arrangement, 3
Antisense codes, 23
Antisense regulation, 172
Apolipoprotein A-1 amyloidosis (FAP type III), 56, 57, 57*t*, 66, 68, 68*t*
Astrocytoma, giant cell (GCA), 217*t*, 218–219, 220, 221
Asymmetric PCR, 33–34
Autonomic neuropathies, in amyloidosis, 64*t*, 65
Autoradiography, 16, 24, 27*f*, 28
   in situ hybridization, 42, 44
   linkage analysis, 40
   mitochondrial restriction sites, 256, 257*f*

base substitutions, 32
end labeling, 20
extension primers, 28–30, 29f
mixed oligonucleotide primed amplification of cDNA, 32
site-directed mutagenesis, 44–45
synthesis of, 23–24
Oogenesis, 253, 258, 265f
Open reading frame, phenylalanine
hydroxylase, 185
Overlapping clones, chromosome walking, 42, 43f
Overlapping elements, 32

Palindromic sequences, 6, 10, 17
Parkinson's disease, 114f, 259
Paternal mitochondria, 246
Paternal origin of mutation, Duchenne
muscular dystrophy gene rearrange-
ment, 95
Paternal transmission
fragile X syndrome, 264, 265f, 267
mitochondrial disorders, 249, 254
neurofibromatosis, peripheral, 161
Pathophysiology. See specific disorders
PCR. See Polymerase chain reaction
Pearson's syndrome, 246t, 255
Penetrance
Charcot-Marie-Tooth disease, 237
fragile X syndrome, 263
Gilles de la Tourette syndrome, 205t
neurofibromatosis
NF1 gene, 161
NF2 gene, 174
tuberous sclerosis, 227–228, 229
Peripheral neurofibromatosis. See
Neurofibromatosis (NF1)
Peripheral neuropathy
in amyloidoses, 59f, 61, 62, 63, 64t,
65, 66
in mitochondrial disorders, 248, 255
PERT87 probes, 76–77, 78, 80f, 86–
87, 87f, 88–89, 100
p53, 171

Phenol-enhanced reassociation kinetics,
76–77
Phenotypes, 5
Charcot-Marie-Tooth disease, 236–237
Duchenne muscular dystrophy
dystrophin analysis, 97–98
severity spectrum, 88–92, 89f, 91t
fragile X syndrome, 265
Gilles de la Tourette syndrome, 201,
202t
HPRT deficiency, 135
Huntington's disease, 116, 121
myotonic dystrophy, 154
phenylalanine hydroxylase deficiencies, 184, 189–190
Phenotypic methods, HPRT deficiency
carrier detection, 139
Phenylalanine hydroxylase deficiencies.
See Phenylketonuria
Phenylketonuria, 181–197
clinical considerations, 184
enzyme system, 182–184
history, 181–182
molecular genetics, 184–190
cDNA clone isolation, 184–185
gene localization, 185–186
gene structure, 186
haplotypes, 187–189, 192f
linkage disequilibria, 188–189, 190
molecular analysis, 187–189
molecular and clinical heterogeneity, 189–190
regulation of gene, 186–187
RFLPs, 186, 187f
tissue specificity, 186–187
molecular medicine, 191–197
diagnosis, 191–192
prevention, 192–195, 193f
therapy, 195–197, 197f
Phenytoin, 156
Physical maps
fragile X syndrome, 266
Huntington's disease gene, 119f, 120
neurofibromatosis 1 gene, 165–168,
166f, 167f

2961-4
5-30